THE BOY

WHO

REACHED

FOR THE

STARS

THE BOY

WHO

REACHED

FOR THE

STARS

A MEMOIR

ELIO MORILLO

WITH CECILIA MOLINARI

HarperOne

An Imprint of HarperCollins*Publishers*

HarperCollins books may be purchased for educational, business, or sales promotional use. For information, please email the Special Markets Department at SPsales@harpercollins.com.

FIRST EDITION

Designed by Yvonne Chan

Illustrations:
Text break ornament (galaxy) © CulombioArt/Shutterstock
Background for decorative capital letters © Zakharchuk/Shutterstock
Page iii © pixelparticle/Shutterstock
Pages 4, 54, 78, 122, 181 © Akito Studio/Shutterstock
Pages 15, 86, 188 © kosmofish/Shutterstock
Page 28 © davooda/Shutterstock
Page 43 © Nadiinko/Shutterstock
Page 66 © iStar Design/Shutterstock
Page 101 © Skeleton Icon/Shutterstock
Pages 115, 147 © Rvector/Shutterstock
Page 133 © filborg/Shutterstock
Page 165 © IconKitty/Shutterstock

Library of Congress Cataloging-in-Publication Data has been applied for.

ISBN 978-0-06-321431-6

23 24 25 26 27 LBC 5 4 3 2 1

To my mom, for loving me unconditionally . . .
through struggles to the stars.

CONTENTS

CONTENTS

INTRODUCTION

The first time I was betrayed by gravity I was three years old. When the clock on the wall of la escuelita struck 10:00 a.m., I eagerly beelined to the silver slide out on the playground. Gripping the railings, I steadily and fearlessly climbed the ladder's steps. At the top of the platform, I turned my back to the inviting slide that would've carried me safely down to the ground, and instead I faced the open air, readied myself to take my maiden flight, and leapt. I relished one glorious fraction of a second suspended in the warm Ecuadorian air before predictably plummeting face-first to the dirt below. Blood gushed out of a fresh gash on my chin, and before I knew it, my nursery school teacher had plonked me onto the back seat of her car and we rushed to the hospital in Guayaquil. In the emergency room, I was laid on a gurney and wheeled to a nearby station. A doctor seemed to appear out of nowhere, gently covering my face before he began the tedious and arduous work of cleaning out and stitching up the wound of a three-year-old boy who believed he could fly.

Gravity won that day. And by the time I became a systems testbed and operations engineer for NASA's Jet Propulsion Laboratory, gravity had won many other days as well. It won the days in Ecuador when the economy was sinking, and the days of my mom and dad's toxic relationship. It won the days in New York when our family's language barrier and financial adversity seemed insurmountable, and it won the day my school guidance counselor changed my As to Bs to make my accomplishments as an immigrant student seem more "believable." But every day that gravity won was also a day when I witnessed my mom's courage to get back up and carry on. You see, gravity doesn't just keep us down; it can also ground us and shape us—it's the weight of being. Along the way, my mom and I found solace in our family and the kindness of strangers. And eventually, I channeled my mom's courage into my unflagging pursuit of an education that would keep us from becoming just another immigrant family statistic. As I got older, I stumbled upon galvanic mentors who opened my eyes to the realities of space exploration, but it was my mom who instilled in me resilience and perseverance, who grounds me in the mental fortitude that allowed me to get to where I am today.

I went from being a toddler who thought he could fly to an immigrant kid who daydreamed about having his own exosuit and spaceship, to a teenager who knuckled down and reached for the stars, and finally to a man who is a mechanical engineer affectionately known as "the space mechanic." I have worked on equipment that soared through space and is currently on Mars, trying to help us understand whether there may once have been life on a planet other than Earth. The journey has not been without its sacrifices and hardships, including hitting astounding professional milestones in my late twenties only to nosedive into deep burnout territory within that same timeframe for not recognizing when to stop, take a breath, and make room in the vastness of space for the vastness of myself. The good news is, not only am I just getting started but also, in a lot of ways, it feels like *we* are just getting started.

There is still so much more to be done to expand humankind's access to space, and I'm thrilled to be a part of the exploration. Working on the Mars 2020 mission allowed me to contribute a grain of sand to helping our civilization understand our place in the universe. How were we made? When were we made? Why were we made? Are we alone? These are the questions that inspire me to push the limits of space, technology, and myself. Despite the conflicts, the catastrophes, and the endless cycle of negative news we experience regularly, I know that one day we'll be able to look back to this period of history and say, "That was when we started it all; that was when we started to venture to the stars."

Regardless of whether your expedition beyond these pages takes you to Mars, to the depths of yourself, or to anywhere in between or beyond, I hope my story inspires you to launch into the spaces you've never explored so that together we can create a better future for humanity, our planet, and this expansive universe in which we play a very small, and yet profoundly powerful, role.

Per aspera ad astra.
Through hardships to the stars.

CHAPTER 1

LAUNCHPAD

From Ecuador to the Unknown

My jaw dropped when I first caught sight of the massive trucks roaring across the colossal construction site and the towering yellow bulldozers carting ginormous scoops of sandy soil from a ditch to a mountain-high pile. Standing next to my dad, I could barely make out his voice through the cacophony of sounds coming from these metal monsters. As the sun beat down on us from the expanse of blue sky adorned by speckled clouds, my three-year-old eyes zeroed in on each machine—it was as if I were watching my own live version of Transformers. I carefully observed their every move, trying to decipher what made those enormous wheels turn, what caused the engines to rumble, how the man driving the bulldozer commanded the blade to lift. I was hooked. From that day forward, I was enthralled by any machine that had an engine: trucks, cars, planes, trains, rockets.

Each time I was given a toy, I wanted to understand how it worked. I'd carefully assess the parts, unscrewing and removing anything that kept me from seeing the full picture of its engineering. But sometimes the quickest way to learn how something works is to make it not work. Once, frustrated with a Transformer whose pieces were too tightly bound, I threw it out of a second-floor window so that it would break apart and I could figure out how its crane functioned. I rushed down to the front patio and found my mom standing next to the scattered fragments. "Muchacho de mierda," said my mom in that half scolding, half *What am I going to do with you* tone she used each time I pulled one of my Dennis the Menace stunts. I looked up at her and replied with a rebellious laugh. She was right: I was a little shit who got a kick out of defying the adults, just like I had tried to defy gravity and fly. The problem was that this was not my toy to destroy, and when neither I nor Mami was able to fix it, she had to buy a new one so we could return it to my neighbor in one piece. That Transformer was an early realization that actions have consequences and we must correct and learn from our mistakes. While this is true for toys, machines, and rockets, it is especially true for humans.

I don't have many memories with my mom and dad together, but there is one trip the three of us took to Riobamba, to visit their friend's new apartment, that sticks out in my mind. This city, also called the Sultana of the Valley, is surrounded by several volcanoes, including Tungurahua and Chimborazo—also considered the tallest mountain in Ecuador—and is nestled in the Chambo River Valley of the Andes. The trip from Guayaquil to Riobamba took anywhere from eight to ten hours, depending on whether there were landslides blocking the way. Although the road was technically called a carretera, meaning a highway, it was really a narrow two-way route that wound its way up the fog-laden mountain and at times merged into one lane, so we had to take turns allowing oncoming traffic to go by before we could continue. The path was hazardous, with low visibility by day and no streetlamps lighting the way at night, so it was best to travel when the sun was still high in the sky.

My dad knew that road like the back of his hand, but that didn't stop my mom from squirming in her seat each time he hit the gas. "Baja la velocidad," she'd say in her soft voice, pleading with him to slow down, while simultaneously slamming on an invisible brake pedal under her foot, scared stiff each time he got too close to the edge of the road. Rather than fight, he'd just laugh off her reaction and step on the gas to poke fun at her for her back-seat driving. Meanwhile, I just looked out the window in awe of the mountain with its cloud-covered forest looming large on one side of the truck and the steep, menacing precipice on the other. I was living in one of my own made-up Power Ranger stories. I could see the Zords materializing from beyond the cliffs, darting across the treetops and leaping into the air, ready to join up and transform into a Megazord, a humanoid battle robot that would protect us from harm.

Mami hadn't planned on having a second child at forty-one, but then I popped out on April 30, 1993—an accidental pregnancy followed by a complicated birth that nearly killed her—seventeen years after her firstborn son, my brother, Xavier. Mami's first marriage had lasted about a year. Shortly after Xavier was born in 1976, the couple separated, but it took her nine years to finalize the divorce because he didn't want to shell out any child support. For most all of Xavier's childhood she had been a single mother.

Then in 1979, at a quinceañera party for the daughter of a friend, she met my dad. Mami walked into the room with her brown hair in a classic bouffant style, a demure below-the-knee dress gracing her petite figure, and understated yet elegant jewelry topping off her look. My dad's green eyes were immediately drawn to this new presence in the room. He crossed the floor in her direction with his signature mustache and short dark hair, dressed to the nines in a button-down shirt, dark slacks, and a suit jacket, his hallmark gold chains adorning his neck and wrist. With his swagger on full display, this tall man introduced himself to my mom and struck up a conversation, his charm immediately commanding the room. Mami nodded and smiled, and when he mentioned he needed help running his business, she offered to connect him to Guayaquil's superintendent—her father was a politician and member of the city's council. What started as a friendship gradually blossomed into a romantic relationship. They fell into a good rhythm as a couple until, two years into their relationship, he got married to another woman without telling her—the first of many collisions in their time together. Hurt and deceived, she immediately broke up with him.

Yet somehow, wielding his charm, sense of humor, and perseverance, my dad was able to wriggle his way back into her life and, with his empty "I'll do better," "I'll be better," "I don't love her like I love you" promises, managed to convince her to couple up once again despite him being married to another woman. Mami was so busy with her career and raising Xavier as a single mom, she didn't have time to read between the lines or question his genuineness, so she decided to look the other way and give him another chance to prove himself. My dad is the quintessential charming, gregarious, flirty, macho Latino. Wherever he goes, he is the life of the party, loved by all. And like the stereotypical macho Latino, he had several ongoing affairs. When I came into this world, I became one of eight children from four different relationships on his side. Although I do have many half siblings, I don't remember hanging out with any of them—likely because they were fifteen to twenty years older than me—except for my half brother Andrés, who had been born only three months before me. Since we were so close in age, my dad used to pick us up and take us on a lot of outings together, like going to a nearby beach to play and cool off in the water. On one such occasion, I looked up from the structure I was building in the sand, stared out at the ocean, and thought, *Those are the biggest waves in the universe!* When they crashed against the dark grains of the shore, it was almost as if the water had veins. That image will forever be etched in my mind as the first time I had any kind of realization of how powerful the forces of nature are. By looking out at that seemingly endless beast of an ocean, I was becoming aware of Earth's magnitude—it was the dawn of my fascination with nature and space, which years later would take me on an unimaginable adventure exploring worlds beyond our planet.

I never really got to know the rest of my dad's side of my family in depth as a kid either. My dad's mother lived in Galapagos, a two-hour flight from Guayaquil. He would visit her often, so I remember feeling her presence through the food she sent us via him. Since she managed a farm, we'd always receive cartons of eggs, chickens, fresh cheese, and sometimes even a six-foot rack of fresh crabs, which my dad would share with our family. He loved inviting everyone over for a barbecue—he'd cook up the crabs and some meat, and had homemade chimichurri at the ready to give everything an extra kick of flavor. I know from stories that my grandmother flew over a few times, and I do remember her calling me Elito, but that's about it. Growing up, I was told I resembled her a lot, from my black hair to my bushy eyebrows, something I saw firsthand when I reconnected with her sixteen years later.

On the other hand, I knew my mom's side of the family well. With my birth, Mami suddenly found herself juggling a newborn as well as a demanding career as a principal of one of the largest private schools in Guayaquil. She had worked her way up after teaching for more than twenty years at two other schools. That's when my great-grandmother—or, as I called her, Abuelita—stepped up to the plate to help my mom. She moved in with us and readily began to look after me while my mom took care of business, which, as a self-described workaholic, had no start or end time.

Abuelita, who was in her late seventies at the time, looked like a cartoon grandmother: short, wavy white hair, greenish eyes framed by big 1970s glasses, and a housecoat snug around her petite figure. There was also Vicky, our beloved housekeeper, who cooked and cleaned—a typical setup for middle to high socioeconomic families in Latin America. She became an even bigger source of support to my abuelita when I learned how to crawl and began to swiftly zigzag around the house, eager to explore every corner of undiscovered terrain.

Every weekday Abuelita would perch me in front of the TV at lunchtime with a bowl of food. I specifically remember watching *Los Picapiedras* (*The Flintstones*) and eating a typical Ecuadorian dish called seco, a chicken stew with rice. (Legend has it the name comes from British oil company employees camped out in Santa Elena in the early twentieth century who kept asking for "seconds.") That's probably also when I started thinking that I could fly like my revered onscreen superheroes Batman and Superman. Since there were no other toddlers in the house, those cartoons and my abuelita's company were my sole entertainment until I took my first steps.

As soon as I turned two years old, my mom began to drop me off at la escuelita, a nursery school run by one of her friends—the scene of my unsuccessful flight attempt. Outgoing and curious, I befriended the other kids right away, and we spent our days playing, and learning how to swim and write our names. That's also where my cheeky, defiant side began to flourish. One time, during a school event celebrating Las fiestas octubrinas de Guayaquil, dressed in the city's light-blue and white colors, my classmates and I were each given a balloon tied to a long, thin stick as a performance prop. As soon as we hit the stage to sing our song in celebration of the city, I grabbed the pointy end of my stick and started popping the other kids' balloons. The teachers tried to stop me, but the crowd burst out laughing as I gleamed from all the attention.

In those early years I was shielded from the problems our family faced. Since my dad was the eldest son, he had inherited his dad's—my namesake's—hotels and some other properties when he died in the 1970s. Everything seemed fine until Mami noticed that my dad's occasional visits to the casino started to become more frequent. Their finances were never tied—he did with his money what he pleased, and she remained financially independent with her own hard-earned salary and savings. Yet she couldn't help worrying about him as she observed his inheritance slowly slipping through his fingers. I vaguely remember Dad taking me to one of the hotels located in Guayaquil's center, but all that comes to mind are musical instruments scattered around the office and a Super Nintendo that I begged my dad to connect for me while he worked. He never connected it, and I never saw any other family hotels or properties after that because my dad gambled them all away. Soon even his construction business went down the betting drain.

As if his gambling problem wasn't unsettling enough, my mom also heard through friends that he was running around town with a new woman. Even though he said that Mami was the love of his life, he continued to cheat on her—I guess a leopard can't change its spots, as the saying goes.

While these events unfolded at home, Ecuador was raging toward what became its major political and economic crisis of the late 1990s. Four presidents took office in the span of four years, beginning with Abdalá Bucaram in 1996, who was dubbed "El Loco" because of his erratic behavior. Within the first six months of his presidential term, the National Congress declared him mentally unfit to rule and removed him from office in February 1997, just at the cusp of a banking crisis that would lead to a spiraling economy. These events increased my mom's unease regarding our future, and it didn't take long before she considered leaving the country.

Mami's parents had moved to New York a few years earlier to live with her sister, my tía Pilar, who had petitioned for my grandparents' green cards. This meant, if we moved there, we'd still be near family. But when she floated the idea of leaving by her close friends and co-workers, they thought she had lost her marbles. "You're going to have to start from scratch. You don't even speak the language. Don't go," they implored. Mami had an extensive network of friends, an active social life, and a career in education she'd worked hard to develop over the years. For all they knew, she was in a good place. In retrospect, I understand her co-workers' disbelief. We lived a middle-class life in a small house within a relatively safe neighborhood. Why leave it all behind?

What no one understood was that, aside from the country's political and economic factors influencing her decision, she desperately needed to get away from my dad. In her eyes, a move was the only way she could end that relationship once and for all. As much as she loved him, deep down she knew they had a toxic relationship that would likely eventually do her and me more harm than good. On the outside, things appeared to be proceeding smoothly, but Mami was the director of flight operations inside this mission control center, and she knew that it was only a matter of time before it all imploded. Emotions were running high, and a turn for the worse was on the horizon. This was her now-or-never moment, and she opted for now.

Our story is not unlike many other immigrants who have had to face their own now-or-never scenarios, oftentimes under much more dire circumstances than our own. However, we didn't have to trek through several countries to cross the border into the United States and seek asylum. For many, enduring a life-threatening journey to a place that may provide safety, stability, and opportunity is a far better choice than having to stay in their hometowns and face economic meltdown, violence, and even death.

When a spacecraft is launched from Earth, its forward velocity combined with Earth's gravitational pull cause it to travel in a curved path. As it gets close to Mars, the red planet's gravity well also affects the path of the spacecraft. The whole process is akin to that of a quarterback throwing a football to a receiver. The quarterback throws the ball downfield of the receiver, taking advantage of the receiver's velocity and direction toward a target. The receiver runs in the same direction as the ball, aiming to perfectly align their hands with the ball in the right spot to make the catch. In a well-executed pass, the receiver's speed and direction will get them to the exact location where they will be able to catch the pass. That's what happens when we launch a spacecraft to Mars. With current rockets, there is only a monthlong window to launch them to the red planet. The difference in elliptical orbits, as well as slightly misaligned orbital planes, cause Mars and Earth to get close to each other approximately every twenty-six months. That means the monthlong window to launch and achieve precise planetary entry, while considering mass limits and landing requirements, happens every twenty-six months. At NASA, we were scheduled to launch our Mars 2020 mission on July 17, 2020, and then the date got pushed to July 30, the very tail end of our window. There was no more wiggle room after that. If we missed that date, our Mars 2020 mission would've had to wait until 2022.

Trial and error, sacrifice, and a dash of defiance—that's what it takes to get to a place like Mars. And that's what it took for Mami to make the decision that would change the course of our lives. There is a photo in one of our family albums of Mami in a suit dress, standing tall in front of the Colegio Alemán Humboldt de Guayaquil, where she was the principal, with an expression on her face that I'd see throughout my life. It's her fierce "I am not to be messed with" look, the one that shows up when she is determined to get something done—the same look I saw take hold of her each time we faced unforeseen hurdles. Bothered and not quite believing she'd go through with it, my dad kept saying to her, "Why are you saying goodbye to everyone? You know you'll come back here crying in three months." Even the school director thought she'd eventually return, so he decided to hold her position for a year in case she changed her mind.

Although it was one of the hardest decisions my mom had to make in her life, she knew that sometimes the only way to fix an unworkable issue is to start over and build something new. The launch window was set, and missing it was out of the question.

CHAPTER 2

LANDING SITE SELECTION

New York

The first time I ascended into the sweeping blue sky was in the summer of 1997, when Mami and I left Guayaquil for good and moved to New York. I wish I could say I was captivated by the hum of the engines as the nose of the plane turned up to the stars, or the expansive mountain ranges and plots of land and buildings below, but all I recall is the piping hot lasagna and pillowy sponge cake on my tray table and the bag of mismatched LEGOs Mami had packed to keep me out of trouble during the seven-hour journey. I spent a solid part of that flight focused on building the foundations and structure of a house, likely a subconscious expression of the first intrepid move of my life.

That initial landing in New York didn't really cause a seismic shift in my world as a four-year-old. We were warmly greeted with hugs and kisses by Mami's family and driven straight across Brooklyn to our new neighborhood, Bay Ridge, bypassing the Manhattan skyscrapers and any inkling that we had just entered one of the most prominent cities of the world. When we reached the compact, four-story apartment building located steps away from one of the main avenues, we walked up two flights of stairs and into a small three-bedroom apartment, home to my tía Pilar, her husband, Eli, and their two daughters. My grandparents had also been living with Tía Pilar and Eli for a few years, which meant that by the time we arrived, all the rooms were taken. Nevertheless, Tía Pilar graciously offered up their living room and pullout couch as our new sleeping quarters. It was cramped for all of us, but we made do. Tía Pilar was the first one of my mom's family to make a life for herself in this city. She had arrived in her early twenties, not knowing anyone or an ounce of English, escaping a nasty relationship and in search of the "American dream." She worked factory jobs at first, but as her English improved, she landed an office job, and by the time we arrived in 1997, she was a receptionist at a doctor's practice. Having established a new life and family here, a husband and two daughters, her old family was now following in her footsteps.

Before coming to the United States, my grandfather had been a politician, one of Guayaquil's city council members. My grandmother had worked as a caterer. In the States, my aunt had managed to get them both work cleaning offices at the doctor's practice that employed her. And my mom—who, despite being a college graduate and former school principal, didn't have the English skills to back her experience—managed to get a job working on an assembly line at a jewelry factory. Soon after, she found a second post, packaging food at JFK Airport, and also began to clean offices with my grandparents on the side to make ends meet.

With her three jobs, my mom and I saw very little of each other during the week, but she never missed dinner at home. I didn't notice her absence during the day. I was only four years old, caught up in my immediate surroundings and in the attention I received from my grandparents. I felt safe with them. My grandfather would take me on walks around the neighborhood, and when school let out, we'd pick up my cousins and head back to the apartment. When the door swung open, my abuela would be waiting inside with some delicious snack. She quietly observed us as we ate and played, her green eyes creasing at the sides each time she smiled. And when our plates were empty, she eagerly encouraged us to eat more or simply stop and talk about our day. Meanwhile, although my abuelo had a governing presence and seemed like a strict and slightly grumpy old man back then—he will not mince words when he disagrees with what is being said; I think I get that from him—he also could turn on the charm with his jokes and mischievous smile.

Watching my abuelo and abuela throughout my life has shown me the epitome of what love can be. As the eldest of six siblings, my mom remembers how her dad would always jump in and take care of them each time her mom was off to the hospital to give birth. He'd feed and bathe them before going to work, always at the ready to give my abuela a hand. Even now, in their nineties, Abuelo is still looking out for Abuela. I often catch him asking her, "How do you feel? Do you need anything? What can I get you?" My abuela gets annoyed at all the questions and shoos him away, but a few minutes later, she'll turn around and make sure he's got a drink or a plate of food and is looked after in the same way.

When my mom suffered some complications and almost died during childbirth with me, my abuela got on a plane in New York and flew straight over to Guayaquil to be at her bedside. Abuelo had to stay behind working. After two weeks apart, he was missing her so much he asked her to come back immediately. The date was set, the ticket was booked, but a volcano eruption grounded all flights that day. My abuelo became so desperate, thinking she might not come back, he told Tía Pilar, "I'm going to jump from the Brooklyn Bridge if she doesn't return to me." That brief time apart almost threw him out of his orbit and into a black hole of despair.

An orbit is a repeating path of an object in space relative to another. An object with an orbit is called a satellite, be it natural, like our moon, or artificial, like the International Space Station. According to NASA, there are planets, comets, asteroids, and other objects in our solar system that orbit the sun along or close to an imaginary flat surface called an "ecliptic plane." Almost all orbits follow an oval trajectory—an elliptical path—but some orbits may be close to circular, like the planets around the sun.

Low Earth orbit, or LEO, is any orbit around Earth between 100 and 1,200 miles from its surface. For objects to stay in orbit around Earth, they must achieve enough horizontal velocity to create momentum to overcome the pull of Earth's gravity. In LEO, this velocity is typically upwards of 17,000 miles per hour, but as an object moves into farther orbital planes, beyond LEO, its velocity is lower. Rockets must achieve these velocities to inject satellites into their operating orbits. Nonetheless, even upon correct orbital injection, the satellites in LEO are constantly falling (losing altitude) and often must correct their trajectory due to orbital decay. Some satellites, typically used for communications infrastructure and weather observations, operate in a geocentric orbit at about 23,000 miles from the surface of Earth, matching our planet's rotational speed, thus appearing static in the sky to an observer on Earth. Satellite systems are generally designed with a twenty-five-year operational lifetime in mind, meaning that after that time span, they eventually fall back toward Earth and burn up in the atmosphere. This may change in the future as technologies such as in-space refueling continue to increase orbital lifetimes.

My abuela is the moon to my abuelo's Earth. She keeps him steady and helps regulate the ebbs and flows of his life. The gravitational force bringing them together has created an enduring family legacy of love and strength that will soon reach the almighty seventy-fifth wedding anniversary milestone. To this day, they continue to be a guiding force in my life. One day, I too hope to find my own moon and experience a similar everlasting love.

A couple of months into our stay at Tía Pilar and Eli's house, in an effort to give them their living room back and out of respect for their generosity, we started spending our nights with my tía Alba. She is one of my mom's lifelong friends who is like a sister to her, and she happened to live alone in a one-bedroom apartment only a few blocks away. She had a career in banking at the time and later became a real estate agent, so she was always dressed to the nines and exuded great strength and kindness.

The first friend I remember making in the States was Tina, Tía Alba's ninetysomething American neighbor who also lived on her own. Since I was too young to go to kindergarten, I spent my first fall in New York between Tía Pilar's and Tía Alba's homes, and several of those afternoons were with my new BFF, Tina—her family rarely stopped by to see her, so she received me with open arms. She had the quintessential grandma look: petite and fragile, with snow-white hair and a housecoat to round it off. Her style reminded me of my abuelita in Ecuador. Each time I visited, she'd place a plate of cookies on the small wooden table in her eat-in kitchen and we'd play whatever game she had at hand. Sometimes it was cards; other times it was number or drawing games. She also had a mini piggy bank in the shape of a slot machine; I would spend hours inserting coins and pulling on the little lever. During what would have otherwise been long and lonely days, Tina and I provided a sense of comfort and company for each other, despite being separated by decades, culture, and language.

I started picking up some English words as I connected the dots between the TV shows I saw at my tía Pilar's house and the toys with numbers, colors, and other simple words that my cousins and I used to play with. I imitated the sounds I heard and strutted around the apartment pretending I was speaking English even though it was gibberish. But when I was hanging out with Tina, since she didn't know Spanish, I had to step up my game to communicate with her. She became my English conversational partner during those months when school was still beyond my reach, and I soaked up every minute of it, already exhibiting a budding love of learning.

About a year after our arrival, I had reached kindergarten age, and my mom enrolled me in a nearby school that was only a short walk away from where we lived. The school's demographic was a mix of immigrant kids and locals, but since my English was still pretty basic, I was placed in the ESL (English as a Second Language, now known as ELL, English Language Learners) program. From very early on, I made an extra effort to learn the correct pronunciation of each word and say it with the right accent. It wasn't just that I wanted to fit in and assimilate, or that I was a chatterbox and knew this was the tool

I needed to make new friends; I had a nascent drive to continuously improve and excel in whatever I took on. I didn't just want to learn English; I wanted to master it. Ever since then, when I am tasked to learn any type of new skill, I always try my best to know it inside and out before moving on to the next challenge. I've never been one to half-ass anything in my education.

Within a month, I was completely fluent and finally able to comfortably talk with the rest of my classmates and befriend them. On the flip side, now that I knew English, I began to feel really bored in my ESL group, so I quickly became that annoying kid in class who raises his hand and says, "I already know this. Why are we going through this exercise agaaaain?" I became a high achiever and was eager to face new challenges. It wasn't just that I was the son of an educator or that I had a desire to prove myself; my greatest joy came from bettering myself. (I am still my own biggest competitor.) Despite being ready to move on, I was required to finish out the year with the ESL group. From a young age, I had a forward-looking trajectory, which in turn steeped me in frustration when I felt like I was being unfairly held back because of standardized or bureaucratic requirements.

In October 1998, my mom had to travel to Ecuador to sort out her pension paperwork to process her early retirement in Guayaquil and a few other logistics to cement our move. Thankfully my abuelo had already petitioned for our green cards, so we had no legal issues with regard to staying in the United States long-term, which is not often the case for a lot of immigrants. Mami was gone for only three weeks, but I felt like it was three months. I yearned for our weekly pizza outings to the local shop a few blocks away. And I missed seeing her smile before I dozed off at night. To make matters worse, communicating with her long-distance wasn't as easy as it is now. There was no WhatsApp, no FaceTime, no Skype. We relied on good old calling cards, and the international rates weren't cheap. Whenever we did manage to connect, our voices reverberated in the echo chambers of our landlines, eating up precious minutes and demanding an extra degree of patience on both ends. As if waiting that eternity wasn't enough, right before she was due to head back to New York, the volcano Guagua Pichincha blanketed the capital city Quito in ashes with its eruption, bringing everything to a standstill and leaving Mami stranded for a few extra days until, to my great relief, she finally made it back home to me.

Two years after our move to the United States, in the summer of 1999, Tía Alba drove Mami and me to JFK to pick up my dad from the airport. A few weeks earlier he had announced his plan to visit and see me, and although Mami wasn't keen on having him so close, she never wanted to keep me from establishing a connection and a relationship with my father, so she went with the flow. She also knew he had no means to stay in this country long-term, which gave her a sense of relief. As soon as I caught sight of his tall figure at the arrivals gate, I rushed toward him and jumped into his arms. I couldn't contain my excitement: the three of us were together again.

Our days flew by as we wandered around the city, showing Dad Times Square, Rockefeller Center, Wall Street, and I showed off my English while attempting to act as a tour guide. One day we made a stop in Chinatown and my dad bought me the Power Rangers Megazord, a combination of all the main team Rangers' Zords that connect to form one big robot—the ultimate gift for a Power Rangers fan like me.

I spent endless hours taking apart my Megazord, fidgeting with each moving part as I imagined the impending battle the Megazord would have to face as soon as I put it back together again. These toys and cartoons introduced me to ideas of justice, good versus bad, redemption, as well as robots and the use of technology to solve problems. While I reveled in my dad's presence and my new toy, Mami had to deal with another reality—my dad's true intention behind this visit: to take us back to Ecuador. "I'm sorry, but I already quit my job," Mami said to him in reference to the position the school in Guayaquil had been holding for her for a year. "I no longer have work there." Moving back to a country that was in a full-blown political and economic crisis without any job prospects made no sense. Her circumstances put an unexpected wrench in his plan. What he didn't understand was that after she had managed to make a clean break from their fraught relationship, as hard as life in New York may have been, there was no turning back for her.

I was so sheltered from all the problems my mom and dad had that I honestly thought he had come to the city to stay. So when my parents told me he was flying back to Ecuador, I was completely taken aback. "What? When?" I asked, exasperated.

"Today," they replied.

How could they drop this news on me so last-minute? Shattered, I clung to him as he tried to say goodbye, refusing to let him go. With tears streaking his face, my dad finally peeled me from his arms and handed me over to my mom, who was also crying at the sight of my suffering. Back in my tía Alba's car, I bawled all the way home. I was only six years old. I would not see my dad again for another fourteen years.

Immigrants and children of immigrants are constantly dealing with goodbyes. Some of the people in our communities have had to say farewell to their parents, grandparents, children, knowing they might not ever see them again. As I grew older and realized I didn't have the means to truthfully say, "I'll see you later," goodbyes became even more impactful. The heaviness I carried from those accumulated departures only began to lift years later, when I tapped into different sources of income that finally allowed me to say, "See you later," and mean it.

After this traumatic separation, my go-to movie became *The Lion King*. Mami believes I turned to that movie because I could identify with Simba, who loses his dad. She's probably right—I found comfort in that film. In retrospect, I can't believe how noble my mom was throughout the entire ordeal. Regardless of the emotional suffering she had experienced with him, she never held anything against my dad, never villainized him, at least not in front of me. And she always pushed me to stay in touch with my father—even when she had to keep him at bay—buying me calling cards through the years so I could continue to hear his voice. I will forever be grateful to her for allowing me to figure him out in my own time.

With the help of their children, and after several years of saving, my grandparents were finally able to rent their own two-bedroom apartment on the top floor of a three-story walkup. Given the tight living quarters at both Tía Pilar's and Tía Alba's homes, it only made sense for Mami and me to move in with my grandparents, where we could upgrade from the living room to our own shared bedroom. The clincher: our new home was in East New York.

To this day, that area suffers from a high crime rate. You need to know where you're walking, because one wrong turn and you could hit a hot block. My mom, protective and anxious, apprehensively approached these new surroundings by making sure I spent that summer safe and sound inside the apartment. With nowhere to go and nothing to do, I decided to turn the window ledge of my grandparents' bedroom into my play area. If I wasn't building LEGO ships and making up stories with my toys, I was immersed in a 1996 Game Boy Pocket I had inherited from my cousins, playing Pokémon Yellow. I don't think Mami understands what an impact that game she gave me had on my young life. I had to strategically build my team based on the characters' strengths and weaknesses. If one was lacking in a specific area, I learned to choose another one that could complement it—hello, teamwork skills! It all became very algorithmic in my head. I took my time to get to know each Pokémon's story so I could carefully choose which ones to catch. I became so immersed in these stories that my Game Boy Pocket became my own universe, an escape from the reality that was just outside the door in East New York.

We mostly kept to ourselves in this new neighborhood, but Mami's fear began to slowly seep into my psyche. I couldn't laugh it off the way I had seen my dad react back in Ecuador. Before I knew it, I was starting first grade at PS 159. I was no longer required to be in ESL and honestly didn't realize I was a minority in a majority Black school until my grandparents and mom emphasized it with their us-versus-them mentality. Their warnings burrowed into my head: "Ten cuidado en la escuela, Elio." "You are different." "They might be mean to you." It conflicted with my reality, one in which I was certain I got along with everyone. I breezed through those first few months at school, and before the fall semester was over, I was placed in the gifted students' class. Although I loved learning and excelling, my favorite part of school was running around with my classmates, playing games outside during lunchtime.

For a while, I managed to rationalize the dread inflicted by my grandparents' and mom's fears. But with time, fear perpetuated itself through anything and anyone that was unfamiliar. I became afraid of people I didn't know, especially the bigger and tougher-looking kids who walked into the ominous building next to our tiny classroom trailer. Only kindergarten and first grade classes were held in the trailer, so the idea of having to go beyond its familiar walls into second grade sent the first shockwaves of panic and anxiety through my body.

With all immigrants in a land where they haven't yet mastered the language, change often makes us vulnerable to fear. And, in turn, fear of the unknown sometimes cripples our logical thought and holds us back when it nestles in the soul. Oftentimes, that irrational fear pushes us to become insular, to keep to ourselves during our time in that neighborhood, and this behavior can eventually mutate into racism. My mom and grandparents had a legitimate reason to be afraid in East New York because it was a dangerous neighborhood. Unfortunately, they began to associate that danger with Black people. They weren't overtly racist; it was subtle—clutching their belongings when someone approached them or crossing the street when they noticed what they deemed to be a "dangerous-looking" individual. Meanwhile, the Black kids in my school were my classmates and my recess buddies, and I simply dreaded the potential bullies in the big building next to our trailer. Once I hit middle school and high school, the racial barriers within my own family began to break as I brought home a diverse group of friends my family loved and appreciated as if they were an extended part of our family. While of course adults hold a responsibility to investigate their own biases and prejudices, sometimes it takes a child to break through beliefs and barriers to show that creating inclusive circles is where it's at.

The Artemis program, named after Apollo's sister and an ode to the 1960s space program, is a robotic and human exploration program that will land the first woman and the first person of color on the moon in the 2020s, bridging the gender and racial disparity gap within space exploration. These new missions, aiming to explore even more of the lunar surface and establish a long-term presence on our only natural satellite, will require grit, dedication, and collaboration among different countries and across industries, plus they will challenge fear head-on—much like the next step in my journey with Mami.

CHAPTER 3

PAD RECONSTRUCTION
Puerto Rico and the Kindness of Strangers

Two years before I was born, my brother Xavier turned fifteen. He was a shy boy who usually kept to himself and preferred to quietly observe his surroundings (quite the opposite of my animated, spirited, talk-to-anyone personality). Mami's colleagues called him a little gentleman. To celebrate this important birthday, my mom gave him a choice: he could have a big party or spend the summer of 1991 with Tía Pilar in New York. He chose New York. By the end of his visit, blown away by the States, my brother asked Mami if he could stay for the school year to fully experience the city. Through a river of tears, Mami assented, not wanting to deny him such a unique experience. She never imagined one year would turn into six.

Time is relative. It dilates and contracts. We leave one world for the next, orbiting celestial bodies for which days, years, even age is relative. A Mars year is almost two Earth years. A year on Neptune is almost one hundred sixty-five years for us. Nevertheless, time can still feel different for us here on Earth depending on our age and our circumstances. The years dragged on for me as a child, while for Mami they rushed by with frightening speed.

When we arrived in New York, Xavier was already twenty-one. He'd been living alone for a few years, something my mom only came to find out when we got there—no one had mentioned this to her before. Turns out he had overstayed his welcome, so my uncle had asked him to move out. Not one to give up, my brother had found a job through some friends as a night-shift guard and managed to make enough money to rent his own place while also finishing high school, persevering through it all. Without knowing, Xavier became a trailblazer who went through life setting examples of resilience that marked my path.

I don't remember seeing much of Xavier while we were in New York, other than occasional weekend visits. I barely knew him, and while I was a rambunctious little kid, he mainly kept to himself, so there was nothing to bond us—the seventeen-year age difference didn't help either. Then he up and secretly married his girlfriend and moved to Puerto Rico, where they had their first child, so I saw even less of him. When news came that baby number two was on the way, Mami planned a trip for us to visit Xavier and his wife, Ruth, meet the new family members, and lend them a hand. That's how we came to spend the summer of 2000 in La Isla del Encanto.

I had recently turned seven years old when we first set foot on the island. Xavier and Ruth picked us up at the airport and drove us to their home, where Ruth's mom, Noemí, also lived, in Caguas, a city located in the Cordillera Central, or Central Mountain Range, just south of San Juan. As the days chugged along, Mami and I started walking this new neighborhood like we used to do in Bay Ridge. She felt much more at ease in Caguas than in East New York—knowing the language probably made a huge difference too. While exploring the main roads and side streets to get our bearings, we came across a private school only blocks from Xavier and Ruth's house. Mami felt an instant pull—and the idea of returning to her teaching roots was reignited. Around a week later, with nothing to lose, she stopped by the school and dropped off her résumé. A few days after that, she received an offer.

In my short life span, I hadn't spent enough time in one place to form deep attachments. The world spun round, its features cloaked in a dizzying blur, with only my mom as the guiding star above me. I had not yet firmly grasped what was always slipping through Mami's fingers: family.

I wish I could say living with my brother finally gave me the chance to bond with him, but although we were under the same roof, we didn't spend much time together. Back then, he had a job that required him to work night shifts, so while I was wide-awake, he was fast asleep. And the weekends were focused on family outings; there was never any one-on-one time, no teaching me how to ride a bike or throw a punch, no stepping in and filling that father-figure role, but I don't blame him. He was much older than me, now a father himself, his plate spilling over with responsibilities. Rather than feeling resentment toward him, I instinctively filled the gaps in our brotherhood with the friends that came into my life.

Simultaneously, I had yet to understand that my parents weren't actually together anymore. Sure, I hadn't seen my dad since his visit to New York a year earlier, but I was used to not having him in my life consistently, so I didn't make much of our circumstances until one afternoon when Mami and I were sitting in my brother's living room watching *Laura en América*—basically a Latina version of *The Jerry Springer Show*. Laura was sitting onstage with a little boy who was around my age and his parents, in front of her boisterous live studio audience, discussing the possibility of their separation. It was one of those episodes you just can't look away from. Suddenly the little boy blurted out, "If my parents get a divorce, I'll kill myself!" The audience gasped.

Without thinking twice, I turned to Mami and exclaimed, "Ay, if you guys get a divorce, I'll also kill myself!"

Mami gasped just like Laura's studio audience. It took her a beat to gather her reeling thoughts, then she carefully replied, "Your dad and I never got married, Elio. And we're not together."

I stared back at her dumbfounded, then fixed my eyes on the TV screen as quiet devastation settled in the pit of my stomach. I didn't get it. How had I not known this sooner? He was the one thing I had been holding on to from Ecuador, and now, with this truth bomb exploded in my face, the family unit I had thought we were suddenly began to slip away, creating a vacancy I hadn't quite experienced until then.

A couple of months after our move to Caguas, I was enrolled in a nearby school and Mami had managed to save enough money to rent a studio on the second floor of a three-unit house located right across the street from Xavier's home. We packed our belongings in our one suitcase, walked across the street and up the stairs, and stepped into the empty makeshift living room area of our new apartment, which had enough space for a small couch and a desk we would later buy at a secondhand store. On the other side of the U-shaped unit was the bedroom area, with two twin beds—hand-me-downs—a small bathroom, and a closet that basically served as the divider between the living room area and our sleeping zone. That was also the spot where my mom took cover each time a thunderous tropical storm rolled through our neighborhood, which was often in Puerto Rico. A few weeks later, centered between the feet of our beds, we placed our most recent thrift-shop acquisitions: a chest of drawers topped with a TV, on which I watched cartoons and Mami caught up on her novelas. As tight as it may sound, this place was a welcome upgrade, and it marked the first time she and I had lived alone together.

We had already gotten to know the neighborhood while living with Xavier and Ruth. There were areas we bypassed when possible or remained extra vigilant in when crossing because getting robbed wasn't unheard of—I saw a man snatch a woman's purse while she was waiting for the bus once—but all in all, I felt safe. People buzzed in and out of the shops in the surrounding blocks, especially on Sundays, when everyone flocked to the city's plaza after church to grab an ice cream at Rex Cream and then took a stroll down El Paseo, a pedestrian walkway flanked by curved lampposts, clothing and shoe stores, knickknack shops, and eateries. I can smell the frituras just thinking about those blocks. Oftentimes, Mami would fuel up with a cafecito while I gulped down a passion fruit juice, then we'd head to the 99-cent store so she could get some school supplies for her students. Another obligatory stop was La Asturiana, a bakery just a couple of streets away that's been open since the early 1900s and sells warm pan sobao fresh from the oven—I can't say I've visited Puerto Rico without buying two of these long, sweet loaves of bread in their signature paper bag. Just add a little butter and you will go straight to heaven.

Without a car at our disposal, having everything from the pharmacy to the grocery store within walking distance made our lives easier. All we needed was our foldable shopping cart and our Dodge

Patas, which is what we called our preferred two-legged mode of transportation. I started taking publicly funded after-school art classes at Programa de Talleres de Bellas Artes, while in second grade, after my new best friend, Jan Josué, and I drew our own comic books with characters inspired by a mix of our favorite cartoons and games. That's how Ray-Man was born. He was a regular dude who discovered a magical staff, and as soon as he touched it, he became a clown-looking figure—somewhat inspired by Sonic the Hedgehog—who beat the bad guys and absorbed their powers (a concept that likely came from the video game *Mega Man*). The illustration bug had bitten, and it was only the beginning.

When I first walked into the studio in the Edificio Víctor Torres Lizardi, my eyes went straight to the blast of colorful student portraits, paintings, and line drawings plastering the walls with the signatures of those who had taken classes here. Then I caught sight of Artemio Rivera, my art teacher, a tall, bandana-wearing, tattooed, white Puerto Rican with green eyes and a rock-and-roll vibe who was a renowned artist and beloved icon in our city. He locked eyes with me, and a smile spread across his face as he waved me inside. I took a seat at the U-shaped table in the middle of the room and scanned the other students, noticing there was a mix of younger kids, teens, and adults. I felt right at home.

"Pick a subject and start drawing," said Artemio. Since this was a multilevel group, there was no real structure in the class. Portraits required too much time and practice for my eager and impatient young mind, so I chose what came easiest to me: landscapes. Eventually, I began to work on an acrylic painting of a forest and a Deku Tree, a pseudo god in *The Legend of Zelda* video game—a piece I would later show at the end-of-year exhibit Artemio organized for us at one of the local galleries. He wandered around the classroom, looking over our shoulders, passionately explaining the specific techniques that would help give our paintings more depth and texture. Any time he recommended a specific brush and noticed I hadn't brought it to the class, he'd pick one up from his desk at the side of the room and hand it over to me, no questions asked. He would never let the lack of a tool interrupt any of his students' artistic flow. Aside from being an inspiring and compassionate teacher, Artemio was also a wonderful and beloved human being. He didn't just teach us how to paint; he told stories and cracked endless jokes. One time, as he was introducing us to a new student, he walked over and stood behind my back and said, "This is Helio," a play on my name with the Spanish word for "helium." "He's the lightest man in the world!" He then placed his big hand on my head and added, "If we don't hold him down, he could float away." As if on cue, when he removed his hand, I slowly stood from my seat, pretending to drift up, and he immediately put his hand back on my head and pushed me down. "You see?" A collective chuckle spread across the room.

He was one of the first men to temporarily fill the father-figure void I had at home. I loved how he lit up any room he walked into with his charisma and charm, like what I remembered of my dad. Artemio didn't just help me fall in love with art; he also taught me how to not take myself too seriously, how to joke around with others while never being mean or making anyone feel inferior.

Unfortunately, classes with Artemio were suspended when he was diagnosed with brain cancer. He bravely ensued a years-long battle against this dreadful disease but ultimately lost. I continued learning about art with Rosa Solivan, one of my mom's colleagues and friends, who taught private classes in her home. I missed Artemio's banter and breezy teaching style, but Rosa quickly found her way into my heart too. I usually got to her place thirty minutes before class started, and she would always greet me with a grilled cheese sandwich and a cup of soda, the perfect midafternoon snack. While Artemio was a free-spirited artist who liked to challenge the status quo and let us paint whatever we wanted, Rosa's style was much more methodical and structured. She'd place fruits on a table and ask us to draw them from our individual perspectives and then we'd paint them. Other times she'd ask us to choose an image from a series and we had to replicate it in another medium. She also taught us how to use different drawing pencils and erasers to create specific shading. The vibe was less joking around and more work oriented in this class. We would spend weeks at a time fine-tuning the same project, sitting in the same seat to keep our perspective consistent, and she was quick to point out any inconsistencies or parts that needed more attention. Although there was a stark contrast between Artemio's and Rosa's artistic methods, I never thought one was better than the other. Rosa pushed us to master each step before moving on to the next, which taught me how to exercise patience and discipline, while Artemio taught us to be present and do our best to enjoy the process.

Many believe engineering is all about logic, but it takes a creative and methodical mind to conceptualize, iterate, and problem-solve highly complex real-world problems. Space engineers in particular are forced to tackle issues that no one has ever faced before, setting precedents for the future while learning from those who came before us. Nowadays, I lean into Rosa's more methodical trains of thought when running through all the different ways a component in a mission can fail, and I use Artemio's free-thinking ways to come up with out-of-the-box solutions to make sure our hardware on Mars lives to see another sol—a Martian day.

In the spring of 2001, less than a year after we began to call Caguas home, my school was holding a Family Day event that served as a chance for students and their families to get to know one another. Since we didn't have a car and the buses didn't run on Sundays, we braved the drizzly morning on foot. As I huddled under Mami's umbrella, occasionally venturing beyond its circumference to splash in the intermittent puddles along the way, a metallic sky-blue Subaru wagon pulled up beside us. The window rolled down to reveal a petite seventysomething lady peering over her glasses beneath a head of dyed brown hair.

"Where are you going in this rain?"

"Buenos días. We're just heading down the road to the school," Mami said politely, somewhat taken aback by this stranger's familiarity.

"I'll take you!" replied the woman enthusiastically.

"No, it's okay, thank you, but we're almost there," replied Mami.

"Please, I insist. There's no need for you two to be in the rain. Hop in!" she said.

Eager for a brief respite from the steady rain, Mami ushered me into the back of the boxy old car and climbed into the passenger's seat. As we drove down the damp streets, the light rain blurring the landscape, Mami and the lady fell into some light chitchat but were in deep conversation by the time we arrived at my school—it was as if they'd known each other for years. As we exited her car and said our goodbyes and thank-yous, the kind woman insisted she would drive us home when we were finished. My mom, moved beyond words, smiled and nodded in acceptance.

Three hours later, we strolled out of the building to find the Subaru waiting for us outside as promised. But instead of taking us home, the lady invited us over for dinner. As she pulled up to her house, we found she lived around the corner from us. Her name was María Flores, but everyone called her Niní. Mami says she was an angel sent from heaven—I couldn't agree more.

Life changed for the better when Niní entered our world. Since she was less than a block away, we'd often drop by—a welcome escape from our cramped quarters. I remember those evening treks—the slight breeze cooling the empty, dimly lit sidewalks; our steps remarkably audible in the quiet of the night, along with the lyrical singsong of the coquí frogs; the warmth radiating from windows that framed scenes of families dining or chatting around a TV. A stark contrast to the booming and ominous East New York streets we wouldn't have been caught dead walking alone on at night. Soon we fell into a nightly routine of watching the evening's telenovela at Niní's before returning to our studio to sleep. On the weekends, she'd break out the saltines and butter, and Lipton soups with an egg on the side, coddling me with her go-to comfort dish. Niní had grandchildren of her own, but they were older and lived far away, so we immediately took to each other and filled that grandmother-grandchild void in both our lives.

She quickly became a mother figure for Mami too. They'd run errands and pick up groceries together, stroll the streets and do some window-shopping arm in arm, and even go along to each other's medical appointments. When Niní's daughter, who suffered a mental health crisis, died of an overdose, Mami was right by her side, comforting her and giving her a shoulder to lean on during such a difficult time. That was my first brush with mental health issues, though I was none the wiser at the time. All I remember is Mami breaking the news to me and immediately segueing into a serious warning: "Always take care of yourself and be careful what you put in your body, Elio." I nodded in response. I quietly observed how the loss had such a deep effect on Niní and her family and felt my heart scrunch up in a ball for her, wishing I could take away her pain. We were family now.

When my mom, in need of supplemental income, began tutoring students at home, we stopped frequenting Niní's house after school and visited on Fridays and the weekends instead, to help her run errands, catch a ride to the supermarket, or drive the thirty minutes to El Viejo San Juan for a laid-back Saturday stroll. At first, I wasn't all that comfortable with those tutoring hours my mom had taken on. Having a slew of random little kids stop by our home between 5:00 and 7:00 p.m. meant I had to quietly stay put in our bedroom area of the studio, which bordered on torture for my rollicking personality. Then Mami spent another hour prepping assignments by hand—we didn't have a computer—for the following school day, which also required some degree of silence from me. She was basically a human version of Microsoft Word.

Soon, two new kids, Benito and Gabriela, started coming over to work with Mami. I remained in my corner as usual, until one afternoon, when Gabriela began yelling, "I don't want to do this!"

"Let's just get this problem over with first and then we'll see what's next," said Mami patiently.

"No, no, I don't want to do any more work!" exclaimed Gabriela.

Oh my God, how annoying! I thought at first, then curiosity got the best of me, and I poked my head around the corner to catch a glimpse of what was happening.

Mami was standing by Gabriela, ignoring her hissy fit and pointing at the page in the book on the kitchen table, calmly asking her to continue completing the exercises. Benito sat on the chair next to his sister, exasperated by her unruly behavior. Having already finished his homework, he looked up and our eyes locked and rolled in camaraderie. Soon I started looking forward to their twice-a-week sessions at our home. Benito and I began to hang out while Mami continued to struggle with Gabriela. They stayed longer than most kids because their parents, Sonia and Robert, often worked long hours. A few weeks in, Sonia asked my mom if she'd let me spend the following Friday night with them. She knew Mami had been doing her a solid by looking after her kids, and she hoped to return the favor. That Friday turned into a weekly event.

Suddenly I went from living in a tiny studio with my mom during the week to spending epic weekends in a sprawling house with a living room, a dining room, a family room, and three bedrooms, one for the parents and one for each kid. They also had three dogs and two cars—one for each parent—their grandmother lived next door in her own house with a backyard, and they all resided in a gated community where no one had to think twice about walking certain blocks because safety was a given. My first thought was *Wait, these are the people I've grown up seeing in the novelas!*

Robert would pick up Benito and Gabriela from school, come get me at home, and then we'd head to Blockbuster (yes, I got to live the last year or two of the Blockbuster experience firsthand), where we'd scan the aisles for the movie and game we'd be watching and playing that night. Dinner was followed by the DVD we'd chosen earlier or whatever show was playing on Cartoon Network. One night we were all settled on the sofas with their three dogs at our feet, watching *Alien Apocalypse*, an absurd sci-fi movie about an alien invasion where giant ant-like aliens imprison humans and force them to work until the humans plan and execute a revolt in the camp. With weapons in hand, they began to decimate the aliens, and green, slimy goo splattered every visible surface. What should have been a terrifyingly gross scene was done with such terrible CGI, we all burst out laughing. Tears rolled down our faces. I held my belly, trying to catch my breath from the nonstop cackling, but was quick to notice, no matter how deeply I tried to inhale, no air seemed to be reaching my lungs. I slapped Robert's arm and pointed at my throat, and they all sprang into action and rushed me to the ER, where I was given steroids to open my airways. I was aware that I had asthma. Mami used to put a nebulizer on my face when I was a baby to make sure I could breathe well through the night—I remember using it as a kid in New York too—but I hadn't had a full-blown attack since I was a toddler, so it wasn't present in my mind. According to the doctors, it wasn't laughing at the damn aliens that had thrown me into this breathing fit; I was allergic to dogs.

Friday night belly laughs sans asthma attacks were usually followed by the highlight of my weekend: Saturday morning cartoons! I saw some of the classics on broadcast TV at home, of course, but since we didn't have cable, I rarely got to see many of the cartoons my classmates talked about at school, until my weekends at Sonia and Robert's house. The one that blew my mind wide-open was *Dexter's Laboratory*. Dexter was a short, young kid with a mop of red hair and thick, black-rimmed glasses, whose usual attire was a white lab coat, black pants, and purple gloves. Unbeknownst to his naïve parents, he had a secret passageway leading down below the house into a massive NASA-like laboratory where he'd come up with cool contraptions and crazy robots, like a talking parrot, that either saved the day or caused mayhem up above. He had a nemesis, Mandark, who was also a boy genius and had a ridiculous evil laugh, and an older sister, Dee Dee, who often managed to blow up his experiments and annoy the hell out of him. My favorite scenes were those that took place in his lab, a metal-laden slice of engineering paradise crammed with equipment, test tubes, and computers. Whenever he was in the process of building his contraptions, my eyes were fixated on the screen. Watching that little cartoon kid in action convinced me that one day I wanted to become an inventor with my own secret lab, where I'd to conduct experiments and build mind-defying machines, especially robots. I really just wanted to be Dexter!

Since a constant stream of kids came over on weekends to play with Benito and Gabriela, Sonia always had a stash of snacks and drinks on hand for us. If we got hungry, she'd heat up some chicken nuggets, bagel bites, mozzarella sticks, and I felt like I had died and gone to junk food heaven. Then we'd sprint to the family room and start playing multiplayer video games, like *Super Smash Bros.* or *Mario Kart*, going head-to-head with each other in sibling-rivalry-type scenarios. I lapped it up. Xavier was too busy being a dad and a husband, and working night shifts to provide for his family, to play around with a little brother closer in age to his children than to him. So Benito and Gabriela quickly became like siblings and their parents like a second family to me. In my mind, this was what it felt like to have brothers and sisters, to have a family that wasn't separated by borders, financial hardship, or divorce.

I became so tight with all of them that when they decided to go on a weekend beach escape to their family-owned villa in Palmas del Mar, Humacao, on the east side of the island, they brought me along without a second thought. This was the first time I experienced anything close to a beach getaway, because Mami and I had never had the possibility to take any sort of vacation. Yet despite living on opposite shores of the financial-means river, Sonia and her husband and kids always made me feel like another member of their family, never an outsider.

They became the people I aspired to be, the kind of family I aspired to have. Their stories—with both parents graduating college and in successful careers—were an affirmation that if I went to school, I could have a shot at a similar life with resources and financial security. But it was also about more than the big house, nice cars, and safe community. It was about having a chance to open my doors and share what I might have with those around me, including neighbors, friends, and family, especially my mom, who no matter where we called home—a bedroom, a studio, or a one-bedroom apartment—always created a soft place for me to land. One day, I wanted to be able to do the same for her.

CHAPTER 4

GROUND SUPPORT
Adverse Weather Conditions

While I remained entranced by school and our newfound chosen family, every day Mami had to continue to prove herself as a teacher. At the crack of dawn, she trudged to her underpaid job at a private school—which did not require teaching credentials but also did not pay industry-standard wages—only to return home to tutor students until sundown to make ends meet. With such a grueling schedule, she became intent on landing a job in the public school system, which would pay her a living wage and better honor her thirty years of experience. To make that happen, she needed Puerto Rico's teacher's certification. The certification classes were held at a nearby university in the evening, so after her final tutoring session at home, she'd pack up and go, dragging me along for the ride.

Meanwhile, I had adjusted effortlessly to our new life in Puerto Rico. When I started second grade, I only knew how to *speak* Spanish. My teacher, pulling me aside, handed me a children's book to gauge my reading level. Taking the time to explain how each letter was pronounced in Spanish first, she asked me to read aloud. Everything simply clicked. Adjusting my pronunciation to sound out the words in Spanish, I read hesitantly but assuredly. Satisfied with my performance, the teacher permitted me to join the rest of my peers in class, though I was kept under observation in case I required extra help, since there was no ESL equivalent here.

My favorite days were when the teacher rolled in the TV and showed us an episode of *NOVA*, a science series on PBS that demystified scientific and technological ideas related to our everyday lives. Even the kids who weren't science fans were drawn in by this docuseries. It has a way of turning ideas and facts into riveting storytelling, creating a deep connection between the viewers and the lives of the people featured in each episode by making their experiences inspiring, personable, and achievable.

I began to further nurture this avid fascination with science, space, and history during my stays with Benito and Gabriela. The HISTORY channel burned itself into my retinas. There were days I went from documentaries on post–World War II to *How It's Made* on the Science Channel, which focused on manufacturing, technology, and space, among other subjects. Anytime a show explained how something was first made and how it evolved throughout the years, I couldn't tear myself away from the TV, especially when it came to fighter jets. The technology that went into building these aircraft sparked a passion in me that would only grow from there. How could the B-2 Spirit bomber—the most expensive aircraft we have to date—fly anywhere, and remain in the air for long periods of time without having to get refueled on the ground? I fell in love with how technically difficult these aircraft were—the aerodynamics, breaking the speed of sound, operating in stealth mode—and I desperately needed to know more. *That's it*, I thought one day toward the end of elementary school. *That's what I want to do. I want to work on those fighter jets or on spacecraft.* I wanted to experience first-hand how things went from a design on a piece of paper to a single prototype, then to mass production. It was a romanticized vision of machines. I only saw their design, abstracting their mechanics from their purpose. Yet, as a child, these shows proved to be my gateway to a world I did not know existed before.

One day Mami and I were in our studio when Xavier and Ruth phoned. I looked over at my mom and saw her grimace as she carefully listened to the muffled voices on the other end of the line. As soon as she hung up, she turned to me and said, "Drop what you're doing. We have to go," and we frantically scrambled across the street. My mom muttered breathlessly, "Noemí just had a heart attack." As soon as we crossed the threshold, Mami went straight to Ruth to see how she was coping with her mom's health crisis and how to help. Everyone else was so preoccupied that I slipped out of the room in search of Noemí. I wanted to see if she was okay. As I entered the bedroom we'd shared with her when we first arrived in Puerto Rico, I found Ruth's mom lying on the floor, her rigid body showing no vital signs. I quickly slipped out of the room, gasping for air. I knew about death through telenovelas and video games, but nothing had prepared me for this scene in real life. The rest of the day was a blur of tears, phone calls, and blank stares. Amid the chaos, Mami managed to quickly drop me off at Niní's to protect me from the aftermath of this sudden death in the family.

That was the first time I remember Mami sheltering me from the storm of life's dramas. The next time I sensed something was off was about a year after Noemí passed away, when Xavier moved into our tiny studio by himself.

Because my mom was so good at safeguarding me, rather than registering the turmoil reverberating within our family, I was more concerned about the nuisance of having my brother living with us in our small space. Then again, maybe I was the nuisance. When Xavier was home from work, he'd sit at a small desk he'd brought with him, hunched over his computer, engrossed in his music collection—what I now know was his escape from the grim reality that was his life at the time. I'd hover around him like a satellite, trying to catch a glimpse of what he was doing on that machine I yearned to learn how to use. Then I'd quietly perch on the small sofa—which was now his bed—behind his desk and peer over his shoulder. As soon as he sensed my breath on his neck, he'd turn around, annoyed, and say, "Stop staring, Elio." I'd withdraw to the other side of the studio, until the pull became too strong to resist, and I'd slide over to the couch again, sneaking another peek while secretly hoping he'd finally teach me how to use the computer.

One night, while the three of us were getting ready for bed, the clatter of frantic footsteps followed by wailing sirens interrupted our nightly routine. A fugitive scaled the flat twenty-foot concrete wall that separated our building from the school next door and landed in our backyard. Cries of "¡Alto! Stop!" pierced our windows. Worry crumpled Mami's face. Then someone bounded up our staircase and slammed their body against our front door in a last-ditch effort to escape. Mami feverishly ushered Xavier and me into the small, narrow closet in the middle of the apartment—her hideout from the island's storms—and slammed the door shut as shots were fired outside. We shoved the shoes aside, ducked down to the closet floor, and covered our heads with our hands. "Shhh, stay quiet!"

pleaded Mami in a frantic whisper. My heart was hammering in my chest, the rhythmic pulse so deafening I feared it would be a dead giveaway to our location. During the minutes that followed, time stood still. We waited with bated breath in that lightless, cave-like structure only to gasp audibly when a heavy knock sounded on our front door. Mami reluctantly hoisted herself up, carefully opened the closet door, and glanced around the apartment. Confirming the coast was clear, she warily cracked the front door open. "Are you okay, señora?" Mami glanced back at us as we made our way to her side and then nodded to the officer. "You're safe now." As he said those words, my eyes floated up to a bullet firmly lodged in our door.

As soon as she received her teacher's certification, landing a coveted position at a public school, Mami's first order of business was to get us the hell out of that bullet-ridden studio.

Two Mars Exploration rovers were sent to Mars in 2003. The prime mission was only meant to be ninety sols (Martian days) but defied all expectations, with *Spirit* making it to six years and *Opportunity* surviving nearly fifteen years. Throughout the development and assembly of the rovers prior to launch there were a slew of issues, including troublesome parachutes that would eventually be used on Mars, mechanical failures, and software glitches. Once the problems were solved, both rovers were launched to Mars three weeks apart, and had six-and-a-half-month trajectories ahead of them. During their long journeys to the red planet, both rovers were hit with some of the largest solar flares NASA had ever observed. These highly energetic particles hit the spacecraft and corrupted the software, which required reboots midflight—a remarkably unusual circumstance that both spacecraft survived. Just a few years earlier, NASA had lost the *Mars Climate Orbiter* due to a unit conversion mistake and the *Mars Polar Lander* due to a software issue. The success of these two rovers was critical for the potential continuation of NASA's robotic exploration of Mars.

After surviving the entry, descent, and landing sequences on opposite sides of Mars, success came hand in hand with new roadblocks that had to be overcome. While exploring Columbia Hills, *Spirit's* right front wheel failed and the team had to come up with a new driving method on the fly. It was agreed that, just like pushing a shopping cart with a stuck wheel, driving *Spirit* backward while dragging the stuck wheel would be the best course of action. After this method was used, *Spirit* had to be parked at an angle to optimize its solar panel energy collection to enable turning on heaters during the cold winter nights, but it survived. Then, during sol 1,226, Mars was hit with a notable global dust storm that put the mission at risk. The rovers were given additional autonomy to power off if their battery charge got too low. Eventually the dust cleared and the rovers pushed through and continued their exploration. Despite the many more scares and challenges to come, and the naysayers, *Spirit* and *Opportunity* endured and went on to have the most successful Martian surface explorations to date, gifting us with images and science measurements that helped us conclude that the Mars surface and subsurface once had plenty of liquid water. Moreover, they left a legacy of inspiration, perseverance, and hope.

For Xavier, one day his saga was over. He moved back in with Ruth and the girls, and his life continued as if it had never been interrupted. Yet a seed of worry had been planted in Mami's mind, watered by her racing thoughts. She fretted about Xavier's well-being, as any mom would, but her anxiety gradually began to build without any of us noticing.

No matter what problems there were with family, with work, with our finances, Mami never complained, but the incessant stream of stress quietly assaulted her every waking hour. Even though her health was under siege, survival mode pushed her to blindly keep moving forward, edging her closer to a precipice. When you've spent your life working tirelessly to achieve a mere morsel of success, an underlying anxiety plagues you, a voice that says, *If you stop now, everything you've worked so hard for could suddenly crumble away and disappear.* This pushes you to keep going no matter what. Mami prioritized my life over hers, until one day her body once again said enough.

"Dora, your face is drooping," said a fellow teacher, overwrought with worry. A clamor arose as bystanders at the school rushed to her aid and called an ambulance. At the hospital, the doctor confirmed her diagnosis: "You had a stroke."

I walked into her room a short while later and froze. Half her face looked like melted butter, and her left arm was lying limp beside her body. Yet no one told me the truth. No one said, "Your mom has had a stroke," and explained what that meant. Instead, the adults told me that the constant change in temperature between the outside heat and the icy-cold air-conditioned classrooms had caused Mami's face to droop. "But don't worry," they quickly added as they sensed my trepidation, "she'll soon be back to her normal self." In the meantime, I would spend the duration of her hospital stay alternating nights between Nini's and Sonia's homes as Mami began her snail-like recovery process.

The stroke was the first time she had a stress-induced health issue. One would think it would've served as a precautionary tale. But as soon as my mom felt better, she fell back into her no-rest-for-the-weary mode. She dedicated herself to other people's problems, accepting any load thrown her way. Asking for help wasn't part of her vocabulary, so she relied on her job for emotional escape—her classic coping mechanism throughout the years. Little did I know that the time would come for me to face the same workaholic tendencies in my own life, a realization that has created a deeper sense of empathy and understanding for her and her circumstances. After all, she was a single mom in her early fifties doing everything in her power to keep us afloat and provide me with food, shelter, and a solid education, while also worrying about my brother. She didn't have many options other than to keep going for her sake and my own. To this day, her need to make sure we all have what we need to thrive continues to override everything else, a switch I hope she can learn how to turn off, especially now that, thanks to her unconditional love, support, and sacrifice, we have more resources. I have come to learn that when we don't listen to our bodies and hit the brakes, our bodies will hit the brakes for us. About a year after her stroke, Mami had a heart attack.

I knew something was off when Sonia picked me up from school. That never happened. Filled with apprehension, I climbed into her car. This time, the adults in my life told me the truth. "Elio, your mom is in the hospital." My chest tightened. "She had a heart attack." My body went numb. "But she's going to be okay." I stared back at Sonia and nodded, wide-eyed, as she broke the news, unable to utter a word in return. Another trip to the hospital, another walk down those glaring hallways, another vision of my mom lying agonizingly still. And then, dissociation. All I remember after that is school, playing games, acting as if nothing had happened. I couldn't handle the thought again of losing my mom, so soon after the stroke.

A few years later, while talking about that difficult moment with my mom, she shared with me that during the week she spent in the hospital, she and Sonia had decided to fill out legal guardian papers for me in case she didn't make it. I was deeply moved and overcome by a powerful sense of gratitude. Sonia already felt like a second mother to me, but this confirmed it. She was willing to take me in as one of her own—I will never forget that.

On January 28, 1986, the Space Shuttle program launched its twenty-fifth mission into the sky: the *Challenger*. It had gathered worldwide attention, as it was carrying a schoolteacher as a tourist passenger. Space tourism was now within the grasp of the public, or so was the perception until that tragic day. At the time of the *Challenger* launch, the Space Shuttle program was criticized for being constantly delayed and severely over budget. Despite the intent of them being reusable, space shuttles required significant refurbishment, and those costs added up quickly. Bringing in citizen personnel to the flights was meant to boost public confidence, primarily showcasing the safety and reliability of the program. So how did such a promising event go so horribly wrong? To start, on the day of that fateful launch, temperatures were in the thirties, below the lowest allowable operating temperature of 41°F. Despite the warnings from low-level engineers, managers decided they needed to meet the current schedule. No further delays would be acceptable. The NASA management team gave a "go for launch," and takeoff occurred despite the red flags.

At barely fourteen kilometers altitude, two layers of O-rings failed in the solid rocket boosters, which caused hot pressurized gas to leak violently, leading to a rupture of the external propellant tank. The spacecraft was thrown sideways, and aerodynamic pressure caused the orbiter to disintegrate on the spot. The solid rocket boosters were destroyed on command by the range safety officer, and the destroyed external tank and remaining debris fell into the Atlantic. It took weeks to find the crew compartment, and it remains unknown whether the crew survived the initial explosion. This disaster is a case study for holding to safety standards above all, especially when human lives are on the line. It proves that speaking up is essential, and sometimes stepping on people's toes may be necessary to get a message across. I can't imagine the despair felt by the teams involved, but I hope that such a tragedy is never repeated within NASA or through any of the new commercial ventures embarking on human exploration of space and commerce.

It took me years to fully grasp the gravity of Mami's health scare. Like the team overseeing *Challenger*'s operations, she didn't heed the warning signs or follow safety protocols. All I knew at the time was that she was finally home, and now she was going to a thing called therapy, but no one explained why, what it was for, or what benefits it could have. Seeking help for mental health issues was still a taboo topic for us; to many it meant that you were "loco," so rather than talk about it, it was swept under the rug as another appointment in Mami's calendar. In retrospect, I am incredibly proud of her for taking such a meaningful step in seeking professional help beyond our circle of friends and family. Going against the grain in a community that still silently frowns upon these actions is far from easy, but she persevered for her sake and mine. At the end of the day, we are not machines, we are not invincible, we cannot handle everything on our own. Although I couldn't rationalize this feeling at the time, this was the beginning of my urgent need to succeed in school, so that one day I could take care of her and give her the break she deserved.

CHAPTER 5

LOAD AND GO

The Academic Dynamics That Shaped My Dreams

By the summer of 2005, I was hell-bent on becoming an engineer, so I set my eyes on finding an environment that valued education and could provide a solid gateway to top US universities. Although my public elementary school years had been awesome, I dreaded attending the middle school assigned to my address. It was known as a hotbed of drugs, violence, and dropouts. And as a kid who loved studying and gaming, drugs, teen pregnancies, and thug culture scared the crap out of me. Even the perreo dance and reggaeton music—which back then was directly related to all things criminal—sent my head spinning. I didn't dare challenge the established order. All I wanted to do was be a model student.

As a teacher and college graduate, Mami knew the right educational path could open doors for her son. Back in third grade, when I brought home my first B, Mami didn't lightheartedly call me "muchacho de mierda," she went straight for the death stare.

"Elio, I expect more of you," she said in her soft yet stern voice.

My heart dropped to my stomach.

"This is unacceptable under my roof. For goodness' sake, you're the son of a teacher. You should know better. Qué vergüenza. This embarrassment is a reflection on me too."

I stared down at my shoes and occasionally glanced up to meet her reproving eyes but said nothing. I knew better than to laugh defiantly let alone talk back.

Then she gave me the final blow. "No Game Boy for a month."

I wanted to protest. Mami never had to nag me to get schoolwork done. This seemed so unfair! I wanted to plead my case to get my sentence reduced, but any form of complaint would've escalated the situation into a spanking. So I bit my tongue, handed over the Game Boy, and took the hit. She had always made it perfectly clear that coming home with anything other than an A in her household was unacceptable. Now I knew she meant it. From then on, I strived to never fall below an A again, and my next B wouldn't arrive for more than a decade.

I continued to spend my weekends at Sonia and Robert's house, and those shows I watched on the HISTORY channel only inspired me to do better in school so that one day I could become one of those engineers. Mami was thrilled to discover that the principal at one of the schools she worked at knew the director of the Colegio Católico Notre Dame. Because of how much he regarded Mami, he was willing to put in a good word for me, but I would have to do the rest myself.

Colegio Católico Notre Dame is an elite private school, a tried-and-true gateway into some of the best universities in the United States. The director said I'd be considered so long as I had high grades. *Done!* To afford the education, we were then offered a reduced monthly tuition, a sum my mom could manage. So I was officially *in*! Relief washed over me. The idea of going to the local middle school had rocketed my anxiety to levels I hadn't felt since I thought I would have to go to second grade in the building with the bigger kids in East New York.

Once everything was sorted out and I was enrolled, Mami took it a step further and brought my friend Jan Josué to this director's attention, explaining what a good student he was and how he deserved a shot too. They accepted him! We were over the moon, because that meant we could continue to be together and help each other navigate this new private-school world that was very much foreign to us.

And so began seventh grade, the year when I began to chart my future. Although I was armed with academic confidence—verging on lofty self-assurance—and couldn't wait to face the educational rigors to come, the first day at Notre Dame was an ostentatious culture shock. I walked past a long line of luxury vehicles pulling into the school's driveway during drop-off with my makeshift rolling backpack in tow (I had basically grabbed my old elementary school backpack, which the mayor had given us public school kids the previous year, and attached it with a pair of bungee cords to a small folding hand truck Mami had bought me) and immediately noticed all the kids wandering inside with fancy bags that were beyond my reach. I remember overhearing some comments at first: "Ay, he's too weak to carry his own books." "Get a bag already." Rolling backpacks were a thing of the past in middle school, totally uncool, but the usual stack of books was too heavy for my skinny body to carry. Backaches had already started plaguing me by sixth grade, so practicality trumped fashion in my eyes. Plus, we couldn't afford the luxury of a real rolling backpack. Logic outweighed any shame thrown my way. As I headed to my first class of the day, I noticed several kids falling into long embraces with one another and catching up on their summer adventures. Many of them had attended the same elementary school, so they had friends here and groups were already formed. I only knew Jan Josué, but thankfully I didn't struggle with befriending new people, and soon enough I had my own crew, many of whom I met through my first new middle school friend, Mara, and in my advanced placement classes.

One of the only times I felt excluded was when well-off kids sported their gadgets as status symbols. It seemed like everyone around me flaunted a brand-new Motorola Razr, while I just had a Virgin Mobile pay-as-you-go phone with the equivalent of twenty bucks on it, which Mami had gotten me for my safety and her peace of mind. Practicality was a necessity; popularity was not. I didn't make much of it until school started and I realized that phones were more than devices used to call or text or play chess. We were on the cusp of social media becoming a mainstream form of communication, and we didn't know it, but the equivalent of likes and comments were already starting to appear. With the Razr, my classmates could easily take pictures and videos and share them with other Razrs via Bluetooth. A rush of dopamine coursed through class when this happened. "Did you see the video Carlos is sharing?" "Did you see the photo of us from the pep rally?" There were group chats, group events, and I would hear all about them by word of mouth. FOMO hit me hard that year.

For those of us who didn't have access to the latest tech—the new Xbox or PlayStation or iPod—feeling left out was somewhat of a given, and it was taxing. I couldn't keep up with game storylines I never played, so it often became difficult to socialize with a particular group of people, which was hard at that age. I wanted to be liked. I wanted to fit in. I wanted to be noticed by the cute girls. I wanted to be invited to parties and events and sports games. But a lot of that was out of my reach. Even though it kept me grounded in my socioeconomic status, I never complained about it to Mami, or anyone for that matter. I had studied hard to earn a spot in this school, and Mami worked tirelessly to provide me with this opportunity, so whining about not having the latest gadget didn't feel right. If anything, it pushed me to focus on my path and continue to excel in the one main equalizer at school: the classroom. That was the one place where we all had access to the same books, assignments, and teachers. We took the same classes. We were on the same playing field—no new gadget could make someone study harder or pass a subject. I wasn't focused on the now; I was focused on the future. I knew that a higher education would eventually open the same doors my classmates already had access to, and that would be my shot at leveling up my socioeconomic status later in life. As the months progressed, I kicked their asses academically and forged friendships in this environment with classmates who are still my ride-or-die people to this day.

After adjusting to the initial culture shock and finding my people at school, I was genuinely happy at Notre Dame. It was the first time I really felt challenged in school at an academic level. Some of the teachers were tough on us, but I used that to build up my skills so I could make it to a big-name university with the rest of my classmates and become an engineer. I was in my element. I belonged.

During one of my recent trips to Puerto Rico, I met up with one of my old elementary school classmates who used to live in the nearby projects. While I went off to Notre Dame, he had no choice but to attend the dreaded public middle school I had wanted to avoid at all cost. While I was deeply immersed in learning and aspiring to achieve more, he was unwillingly thrown into a world of drug lords, high addiction rates, and kids dying on the streets in pursuit of turf control. Not belonging in these circumstances didn't just lead to feeling excluded or being bullied; it also could lead to death. He fell victim to this harrowing environment and lost his way for a while but had since managed to summon the strength and resilience to pull himself out of that dark hole and forge a better path forward, with no means, no help, no support. Now he has a beautiful family, peace, and a stable job as a mechanic that he loves. But many others in his class weren't so lucky. Only a handful of my elementary school friends and I managed to obtain college degrees and the careers we had dreamed of as kids, but it wasn't simply because of our academic prowess—we all had help along the way.

During the summer of 2006, between seventh and eighth grade, I signed up to participate in a summer camp hosted by Notre Dame where we taught arts and crafts to younger children who lived in the local residenciales or caseríos. I had gone to elementary school with kids from the projects and remember staring out the school bus window at some of the people on street corners, wondering, *Why are they lying on the pavement?* Mami later explained they were likely drug addicts. There was fear in her voice as a warning followed: "You should never be like them." That was the first time I was introduced to this concept. In retrospect, I wish it had been handled differently. These people had been failed by society. Instead of being scared of becoming like them, we should do something to help both them and their families. This summer camp was my first step in that direction.

I knew the limitations these kids suffered, like not being able to freely play outside due to safety concerns. I'd had a taste of that back in East New York. Although I went to Notre Dame, I was still in a low socioeconomic group, and on many levels, I could honestly relate to the kids in the summer camp more than I could relate to some of my classmates. When the camp day was over, my classmates got picked up by their parents in Lexuses and BMWs, and I walked home. After all, I lived only a few minutes away from the caseríos. As a result of all this, I was eager to be there, to help my people, to give these kids a respite from their hard lives and the responsibilities they had to carry at such an early age. I wanted to show them what was available beyond those walls, those blocks. I wanted to help them tap into their creativity and maybe even forget about everything for a few hours and have some fun just being kids.

As those long summer days rolled by, I saw some of these fearful and withdrawn children blossom. Their brows unfurrowed and gave way to wide smiles after finishing an art or craft project, a sense of accomplishment that they were later able to take home and share with their families. A positive beam of light in their gloomy circumstances. That's when I truly began to discover the power of giving back. One act of kindness can change someone's day. One helping hand can change someone's attitude. One outreach program can change the course of someone's life.

The first mission solely dedicated to studying methods for the defense of Earth from an asteroid or comet is the Double Asteroid Redirection Test (DART). This mission, launched on November 24, 2021, was designed to study the Didymos asteroid and its asteroid moonlet Dimorphos. DART successfully slammed into its target, Dimorphos, on September 26, 2022, and the footage acquired returned images like nothing we'd ever seen before. At the time of this writing, the LICIACube companion spacecraft trailing DART and managed by the Italian Space Agency, has yet to finish downlinking images that will provide additional details of the impact. Conclusive evidence of any change of velocity of Dimorphos, which is what is needed to divert a trajectory, will be further mission success. The goal is to have an effective method ready to go should we detect an asteroid or comet on a collision course with Earth. If this velocity diversion is concluded, then a DART-like mission may be added to our toolbox to protect the planet.

Missions like DART are a reminder that international collaboration with complex space operations adds incredible value to the global well-being and safety of our civilization. If more nations collaborate in the future, further initiatives will continue to break barriers in geopolitical tensions.

When Xavier and Ruth announced they were moving to Florida, Mami was caught off guard. They were the reason we had come to Puerto Rico, and with them leaving, something began to shift in Mami. We moved into Xavier and Ruth's Puerto Rican house after they relocated and went from finicky dial-up internet to the smooth-running beauty of broadband, which allowed me to seamlessly keep in touch with my friends via MSN Messenger and play *RuneScape*, an interactive quest-based game that takes place in a magical world where your knight avatar goes against dragons and crazy creatures and collects valuable items. The innovative thing was that my friends and I could play together and chat through the game, making it a social tool in our group. It also showed that technology, like education, has the power to level certain playing fields in life—another reason making technology more accessible to everyone is imperative.

In those six years, Puerto Rico had become my home. I even had my first crush. Andrea and I had become friends back in seventh grade; we had the same classes, had lunch together, and hung out a lot, but I had felt nothing more for her until I saw her again in the fall of 2006. I didn't know what the heck love was at the time. I was only thirteen and had very few examples of romance outside of my grandparents' story and telenovelas. There was just something about her that I found exhilarating, and all I wanted to do was spend more time by her side.

Meanwhile, I was clueless to the ways Mami was becoming untethered to the island. Regardless of her efforts to fit in—entering the public school system with humility despite having reached the rank of principal in Ecuador; adapting as best she could to her new loud, boisterous Puerto Rican friends, who greatly contrasted her more reserved Ecuadorian roots—she still felt like an outsider, and for good reason. As much as I love Puerto Rico, it can be a complicated place for immigrants. Its complex history as an unincorporated territory of the United States inspires nationalistic tendencies; that nationalism is a reaction to the people's circumstances, a way to survive the limbo they live in, where the territory is neither a US state nor an independent country. But it bleeds into everyday life. When the shit hit the fan and my brother ran into some legal troubles—which were not of his own making—Mami's colleagues immediately began to view her as a foreigner, spreading rumors throughout her school that she was involved in some shady business. Nothing could've been further from the truth, but once a rumor mill starts churning, it's hard to stop it.

I didn't really suffer any of this discrimination because I was young enough to have assimilated the culture and accent, unlike Mami, who clearly stood out as an immigrant with her Ecuadorian cadence, soft mannerisms, and quiet demeanor. It's hard to explain what makes people tick and why they go to such lengths to make someone feel like they don't belong, but it slowly chipped away at my mom's fortitude until one day she'd had enough.

It happened in October 2006. Mami sat me down and said the words I had hoped to never hear again: "Elio, we're leaving." My stomach sank. "Things are taking a turn for the worse and may soon become unsustainable for us. We're moving back to New York. Your uncle has space, so we'll be living with him." As she continued explaining that I would be okay, that we'd find a good school for me to continue growing academically, an overwhelming sense of defeat blurred my senses. I thought we had been putting down roots—I *felt* rooted. I had been thriving at school, had an amazing group of friends, and had a clear vision of what would happen over the next four years. But her statement wiped it all away like a supercell twister. *How could she have made this decision without me?* The lack of control over the situation infuriated me. Suddenly I was facing an unknown future, one that I didn't want, one that I'd never asked for. My thoughts rushed to my friends, my chosen family, my new crush. I was drowning in the what-ifs of it all, hoping it was just a bad dream.

The remaining month took somewhat of a somber turn. My friends were in complete shock when I broke the news of my impending departure. We tried to make the most of what little time we had together, but the time was shrouded in a heavy mist that wouldn't lift. To say our final goodbyes, we decided to meet up at the local yearly holiday fair. Beforehand, I joined three of my closest friends—Pedro, Andrea, and her best friend, Fabiola—at Andrea's house for a bite to eat and some last-minute photos, and then we caught up with the rest of the gang at the fair. Despite the oncoming departure, I had already accepted the outcome and remained present, enjoying every last celebratory minute with my crew.

Later that night, Andrea pulled me aside and gave me a parting gift: a necklace with a small Puerto Rican flag hanging from it like a charm. I wore that thing until the chain broke, and I still have the pendant stashed away for safekeeping. Earlier that day, my Health teacher had everyone pool a few bucks together to give me a Pokémon game. I also received individual letters with sweet messages that I cherish to this day. We exchanged phone numbers and MSN handles and vowed to stay in touch. I didn't realize at the time how our connections could've easily faded without the dawn of social media, but it was already second nature to us. And so, with a heart as heavy as lead and tears in my eyes, I said goodbye to my first meaningful circle of friends.

A day or so before we were due to depart, Jan Josué and his mom came over to our house. His mom gave me a *Final Fantasy* game as a going-away present, which I would've given back in a heartbeat if that meant I could stay in Caguas. And then Jan Josué released the final emotional blow. He approached me, handing over a box that contained all of his Yu-Gi-Oh! cards, the ones we used to spend hours playing with together and at tournaments, the ones we used to obsess over as kids. In giving me his cards, he basically gifted me our childhood.

The following day, December 6, 2006, we sat by the boarding gate at Luis Muñoz Marín International Airport, waiting to catch the flight that would take us back to New York. I looked out a nearby window toward El Yunque National Forest in the distance and began to weep inconsolably, like the torrential rains that fall on my beloved island. As much as I got Mami's reasoning, I couldn't help but feel misunderstood by her. I wasn't angry. I was heartbroken and hurt. I felt like she was clipping my wings.

After years of listening to her preach about the importance of a solid education and a clear path to college, now we were suddenly going back to square one. I no longer had a clear vision of what could come next, and that thrust me into a state of anxiety, distress, and deep anguish. I spent that four-hour flight immersed in my games, unwilling to engage in any conversation with Mami, wallowing in my circumstances like a wounded bird unsure if it will ever be able to ascend and soar again.

CHAPTER 6

DEFYING INITIAL CONDITIONS

Built to Be Tougher than Expected

To date, only thirteen Hispanics or people of Hispanic descent have been to space. They overcame incredible obstacles to achieve what many thought would not be possible from a societal and technical perspective. José Hernández went from working on farms through high school to becoming an astronaut in 2009, after being turned down by the astronaut training program eleven times. Franklin Chang-Díaz became the first Hispanic to fly under the US flag in 1986. Born in 1950 in Costa Rica to a Chinese immigrant father and a Costa Rican mother, he emigrated to the United States to finish his high school education and eventually completed a mechanical engineering degree at the University of Connecticut, followed by a PhD in plasma physics from MIT. Dr. Chang-Diaz has logged more than sixty-six days in the Space Shuttle program and founded the Ad Astra Rocket Company to develop highly efficient electric propulsion engines that may be used in the future for interplanetary travel. And Dr. Ellen Ochoa, in 1993, became the first Hispanic woman to fly in space. She is a third-generation American raised in Southern California. I can't help but imagine the challenges she had to overcome as a woman, especially when she began her career, to establish technical authority throughout her many years in STEM. She is now an expert in optical systems and holds various related patents. She not only is an astronaut but also has an incredible management career and is now serving as the first Hispanic and second woman to lead NASA's Johnson Space Center as lab director.

These are just some brief details of the inspiring stories of known Hispanic space engineers. There are hundreds of us in the industry today, and just like those who came before us, we are making sure to continue to hold the doors wide open for the next generations.

In 2006, we arrived in New York on the cusp of winter, when the long, blustery nights ate away precious hours. My tío Oscar and his wife, Jenny, took us in this time in Mill Basin, Brooklyn. Mami bunked with my tía Miriam, who was also living there at the time, in the spare room of this three-bedroom house. I slept on an inflatable mattress in the living room downstairs with their cocker spaniel, Gringa, where I'd stay up late watching *American Ninja Warrior*, *George Lopez*, and *The Fresh Prince of Bel-Air* on TV until I dozed off—by then, my allergies had subsided, so I could enjoy Gringa's company without having a full-blown asthma attack. When morning came, I would lug the inflated mattress up to Mami's room and leave it there until it was time to call it a night again. As the days slowly ticked by, my lips were permanently chapped by the frosty weather, cracking and bleeding like my heart. I connected with my friends online, but the chats were tinged with knowing we were now more than a thousand miles apart.

With no winter clothes to keep us warm, my uncle gave me a hoodie and soon after my aunt bought me a green Lands' End jacket that I carried with me for several years. About a week or so after our arrival, my tía Pilar swung by to pick me up and innocently took me to see Santa Claus at a mall, as if middle school boys loved Santa. I realized then that she still saw me as the same little kid from seven years ago. We had a lot of readjusting and reacquainting ahead of us.

I still lived in Spanish in my head, often catching myself saying "Con permiso" in a store instead of "Excuse me." I bit my tongue to resist greeting people on the street. The hugs acquaintances offered instead of a quick kiss on the cheek threw me off. I missed my island.

Soy Boricua aunque naciera en la luna. Although born in Ecuador, I'm a Boricua through and through. I speak Spanish like a Puerto Rican and know where to go for the best pan sobao in the area. Suddenly I was no longer perceived as simply "Elio." I was different. I was a foreigner. I was an immigrant. Was this what my mom had always felt? Maybe now it was my turn to face a round in the ongoing battle of how others perceive me versus who I am.

Sullen and listless, I was little help, so my mom led the effort to find my next school. Conscious of Notre Dame's prestige, we first tried to find a comparable educational haven. But when we heard the private school tuition was in the tens of thousands of dollars and there were no scholarships offered, we quickly scratched that idea off our list. Needing a quick solution, Mami decided to enroll me at the local public school—Roy H. Mann IS 78—so I could finish eighth grade and we could use those months to figure out where I would transfer to for high school.

We visited Roy H. Mann a few days prior to the start of the spring semester. Mami and I carefully explained my situation, the kind of school I had come from, the academic rigor I was used to, yet none of it stuck or seemed to matter. They blindly decided to place me in the middle tier of their three-tier structure. In each grade, top students were in the top tier, which in this school was composed of mostly white kids; the middle tier was for those who didn't quite make it into the top but were better than average, and it held a more diverse pool of students; and the bottom tier was where most of the Black kids were placed and remained regardless of their merit. Within each tier there were top and bottom groups. I was placed at the bottom of the middle tier. "It just wouldn't be fair to the other students who have worked so hard if we placed you in the first tier from the start," we were told at the school office. My qualifications were moot. I was back to square one: the new kid on the block, the one who others saw as a foreigner from a Spanish-speaking country. I spoke English fluently and was coming from Puerto Rico, a US territory, so why was I not worthy of the same opportunities the local top students had? Why was this school able to undermine all of my previous hard work and academic success?

On my first day in class, I immediately felt out of place. Students didn't pay attention, didn't seem to care if they turned in assignments on time, or at all, and constantly interrupted and disrespected the teachers. I looked on dumbfounded. Back in Puerto Rico, we could talk smack and be a little disruptive at times, but there was a line we knew not to cross, ever. That line did not exist here. Our Math teacher straight-up burst into tears after losing control of her wilding students one afternoon. That teaching position had a quick turnover.

While math came way too easy for me, I struggled to catch up in all my other classes, and quickly realized that Earth Science would be the biggest challenge of them all, and not in a good way. They had already covered topics in the previous semester that were meant to be built on in this semester, and without that knowledge, I was quite lost. As the hours passed, I dragged my feet to each class, my frustration intensifying, and by day's end, as soon as I saw my mom waiting for me outside, I erupted in rivers of tears. Noticing my distress, the Earth Science teacher, a Russian immigrant, walked over to us and said, "I understand. I also had to leave my country when I was about your age and remember how devastated I was too." I nodded appreciatively but remained inconsolable. Mami quietly mouthed, "Thank you," as we turned to amble home together, side by side.

In the early 1900s, people had a notion that there may be an alien civilization on Mars that would ping us through the newly widespread radios. After looking at the planet through telescopes, astronomers noted that was far from the truth, but they started to hypothesize that water must once have been present due to the features that resembled our own lakes and rivers here on Earth. NASA eventually sent orbiting and flyby spacecraft, the Mariners, in the 1960s, which gave us our first real photos of Mars. By 1976, NASA landed *Viking 1* and *Viking 2* on the surface, which performed some basic experiments to determine if life existed on the Mars surface—their results were inconclusive, but scientists knew this wasn't the end of the road. It took more than twenty-five years for NASA to get back on the surface of Mars with the *Pathfinder* base station, which landed the *Sojourner* rover in 1997, named after US abolitionist and women's rights activist Sojourner Truth. *Sojourner* helped NASA develop rover operations methods for Mars. It also imaged round pebbles that suggested water had once flowed there, determined the exact pole of rotation of Mars with radio tracking, imaged ice clouds in the early morning, and characterized temperature fluctuations in the atmosphere. Every mission on Mars has helped NASA and the scientific community understand the red planet progressively better, just like every educational experience has helped forge my path forward with growing empathy and understanding for myself and for people from different walks of life.

My tears became a fixture in the weeks that followed. The trigger: planes. I'd sit at my desk by the window and stare up at the sky, wishing, praying this was just a nightmare, and each time I saw a plane fly by, I'd burst into a sob. That metal machine in the sky had the power to pluck me from this torment and return me to my friends, my old classmates and teachers, my real life. Yet with each passing day, that old life faded into the shadow of my new one.

One day some of the kids teased me for crying in class, for being foreign, but I just glared at them, and yelled back, "What are you doing?" Their smirks faded and the room went quiet. "Dude, yeah, I want to go back," I added forcefully. Aware that my suffering went beyond any reaction their snarky comments could incite, after that day, they left me alone.

Although Mill Basin is a well-off neighborhood, many kids who went to my school were bussed in from Flatbush and nearby projects. A lot of them didn't have a supportive environment outside of these four walls. Life had torn down their innocence, pushing thirteen-year-olds to act like grownups. They came to school bragging about their gangster way of life, the guns they handled, the girls they banged or knocked up, and my naïve jaw dropped in disbelief. *What was happening? We're only thirteen!* This place was another level—a blow to the gut and a reality check, to say the least. I was no longer in a Catholic middle school bubble in Puerto Rico; I was experiencing the crude streets of New York City within my own classroom. I needed to toughen up.

As the days dragged on, I searched the school for potential friends and was immediately drawn to some other outcasts. Alex was Puerto Rican but he had grown up in Brooklyn. He suffered bullying at school because he had a small arm that hadn't fully developed, but he put on a brave face and attempted to let what others said slide off his back. He was one of the first kids I had lunch with, and eventually we became buddies. Then Clifford and Jeffrey, nerdy Black kids, started joining us at lunch. I bonded with a few other students too—like Adisa, who was born in Guyana to immigrant parents but had moved to the United States with them as a little kid, kind of like me. But Alex, Clifford, and Jeffrey became my main crew that semester. And then there were Nikita, a Russian kid, and Michael, an Italian American boy. They taught me how to play handball, and I had fun with them for a while during recess until I began to realize the things they dared me to say or do, which I innocently went along with, were really unflattering and straight-up stupid. Like the time they convinced me to use the N-word when I was completely unaware of the weight it carried. Since they and other kids who weren't Black used it so freely at the time, I thought it was okay, but it only revealed a clear gap in my understanding of racism and social justice, issues that rarely had been discussed in Puerto Rico back then. After one too many laughs on my account, I began to wake up to their antics and kept them at arm's length from then on.

The other thing that set me apart from the rest of my classmates was my sense of style, or lack of one. I had spent the last six years of my school life wearing a uniform five days a week, which meant I didn't have to worry about clothes, and I didn't give it much thought now either. Plus, since I was sleeping in my uncle's living room, there wasn't much space for me to keep anything that resembled a varied wardrobe. I started school with a couple of hoodies, two pairs of pants Mami and I had scored at a local thrift shop, the green Lands' End jacket my tía had given me, which served its purpose of keeping me warm, and one or two pairs of shoes. I didn't give any of this a second thought until some little wannabe sneaker-head punk started taunting me one day. "You only have one pair of shoes!" he exclaimed in front of the class after noticing I always wore the same ones.

"Yeah, so what? What's your problem?" I yelled back. Thankfully I didn't understand the sneaker culture, so I was able to brush it off. They were preoccupied with this status symbol, while all I cared about was doing well in school. I accepted my otherness. I knew I was different in more ways than one, especially when it came to academics. And this arrogance allowed me to push through and keep my eyes on equalizing the field in my future through my grades today.

One morning a classmate noticed the necklace with the Puerto Rican flag that Andrea had given me, the one that faithfully hung around my neck. He walked over to my desk and started up with me. "I'll buy it from you."

"No," I answered.

"What if I offer you more money?" he said, purposefully riling me up.

"It's not going to happen."

"I'll give you my PSP for it."

"Dude, absolutely not. This is not for sale!"

I missed my neighborhood. I missed my friends. I missed Andrea. I missed my weekends at Sonia's house. I missed hanging out with Niní. That Puerto Rican flag kept me connected to all of them.

During the last period we had Art with my favorite teacher at IS 78, a middle-aged white woman with short curly hair and glasses. She noticed I was a wreck emotionally but also that I was bright, despite my struggles to catch up in my current classes. She was quick to understand that I didn't belong in the tier the school had placed me in. So on the days I stayed behind to help her clean up, she'd talk to me about the different high schools I should consider and inadvertently became a far better adviser than the actual counselor I had been assigned.

Since I was already late in the New York City high school application race, I had started working with my counselor, a Black Caribbean man, early on in the semester. Like with college, students began their application process for specialized schools in the fall and usually received their acceptance letters in the winter, but I had missed this crucial window of opportunity. That left me with the "supplemental list of schools," the leftovers, the ones that still had a few spots to fill. Not really knowing where to begin, I decided to filter the possibilities based on their names; if it had "technical" or "aviation" in its name or sounded like a STEM school in any other way, then it made the cut. But I didn't know exactly what to look for or what questions to ask—like if it offered AP classes, if it had a college pipeline process, if it had a robotics program. Other than the guidance from my art teacher, I was pretty much flying blind. Eventually, I whittled my list down to five schools that I sensed would be a solid match for me, the ones that would keep me on the path I had started so I could join my old Notre Dame classmates at one of the big-name universities we had dreamed of attending together and finally become an engineer. The change and turmoil of moving and starting a new school meant that I had to pivot my strategies, but it never deterred me from my ultimate career goal. I was hell-bent on making it happen one way or another.

My counselor was there to help me with the application process; I figured I could lean on him for guidance to whittle down my list of choices. He explained that some schools required an entrance exam, which I was more than willing to take. One day, as we continued to prepare to submit application materials to get me a spot in one of these schools, my counselor glanced up from the sheet in his hands, his face impassive. "Listen, no one is going to believe that a foreign kid, just off the plane from Puerto Rico, has such good grades. It's going to seem like you lied. So it's better if we bring some of them down to a B," he said casually, as he began lowering my grades.

"But those are not my grades," I insisted. I hadn't received a B since the third grade.

He remained unfazed. "Trust me," he replied in a forced reassuring tone as he handed me that soul-sinking sheet with my tweaked grades, "it's better for you if we leave them like that."

I walked out of that meeting exasperated. My academic success was everything to me, but a part of me also thought, *Maybe he's right, and this is just how it is here.* That evening, as I told my mom the whole story, I could see fire burning bright in her eyes. That fierce expression from the photo I knew so well reemerged and told me this episode was far from over.

The next day, she went straight to this man's office to talk to him, taking my uncle along to help with the language barrier, only to be met with a similar answer: "Don't worry. These grades don't really matter." He had no idea who he was talking to. You can never tell a teacher and former principal that grades don't matter. Realizing she would get nowhere with this man, she left the building and went straight to the school district and sought out someone who spoke Spanish so she could give them a piece of her mind.

"This is absolutely unacceptable! I have been a teacher for thirty years and I would never dare lower anyone's grade for any reason. All I want is for you to fix this error immediately. What's right is right."

My grades were corrected, and her reaction served as a life lesson I hold tight to my heart to this day: we must always stand up and fight for what is right.

Never in my wildest dreams would I have expected my own assigned adviser to steer me in such a wrong direction. Under what framework did he think that was a good idea? This man had the power to influence the lives of hundreds if not thousands of students throughout the years. How many other kids did he do this to? How many students who didn't have a mom like mine—who fought tooth and nail for my academic success—turned to him as their sole guide? How many were screwed out of opportunities that were rightfully theirs, opportunities that could've set them on a life-changing path?

Mami and I woke up early on a cool Saturday morning and headed outside with our printed map to the nearest B100 bus stop, the first leg of our trip to NEST+m, a top-ranking STEM school in the Lower East Side. At Kings Highway station, we hopped on the B train to Broadway–Lafayette Street, then transferred to the F train for three stops, before getting off at East Broadway and walking the remaining fifteen minutes of the hour and a half commute to the corner of

Columbia and East Houston. NEST+m was one of the few schools that required an entrance exam to check students' baseline math and reading comprehension levels. Upon arriving, Mami waited outside while I entered the large, blue-tiled building alongside a cluster of fellow prospective students. As we were guided to the testing room, I struck up a conversation with a tall, thin Russian kid with longish blond hair named Serge. We were the only two kids from Brooklyn that day. Once the test was over, Serge and I walked out of the building, joined Mami, and headed to the train. A few relentlessly long days later, during first period, I noticed the teacher had a stack of envelopes on her desk. Could this be it? I watched as the teacher grabbed the pile and started handing them out to the students. Adisa tore open his envelope and smiled wide when he read he'd been accepted to Midwood, which had the music and arts programs that aligned with his passions. A Middle Eastern classmate, Imani, with whom I'd developed some friendly competition, opened her letter and nodded joyfully as she read the news. Heart pounding like a bass drum in my body, I watched as the teacher approached my desk. I politely grabbed the letter, said, "Thank you," and took a deep breath. After carefully opening the envelope, the first thing I noticed was the NEST+m letterhead followed by the word "welcome" in the opening sentence. I smiled so hard my cheeks were on the verge of cramping up. When I shared the news with Mami later that evening, we both breathed a colossal sigh of relief. Maybe the damage could be undone.

CHAPTER 7

LAUNCH VEHICLE ROLLOUT

Developing an Empire State of Mind

Timing had been in my favor: my enrollment at NEST+m coincided with its initiative to expand its class size. The senior class prior to my freshman year had fifty students. Our incoming class had a little more than a hundred. There were kids from Brooklyn, the Bronx, Chinatown, Washington Heights, Harlem, and the nearby projects, not just affluent Lower Manhattan students as in the past. And it was a big relief. I wasn't the only newbie coming from a different middle school and another borough.

Entering my first class, I was immediately struck by a familiar face: Serge, from the entrance exam! We gravitated toward each other that day and became fast friends. As the week unfolded, more classmates joined our hang and soon we became a diverse group of buddies—from Russia, China, Honduras, Harlem, Puerto Rico—with varied accents and cultures, all under the same roof. I was finally experiencing the definition of and magic that is New York City.

Ninth period was a drag for freshman because it meant we had to stay an extra hour in school while everyone else was done for the day, but it introduced me to a class that surpassed my expectations: physics. My first exposure to the subject spoke to my mechanics-loving mind. It felt easy and intuitive from the start. We explored trajectories, conservation of energy, motion, waves, light, sound—I couldn't get enough of it.

And then there was Kevin. Like me, he was an immigrant, having moved from China when he was twelve. Kevin was the name he had adopted here. But we were an assembly of international friends ourselves, so I asked if I could call him by the name he identified most with. "It's Wei." His ability to overcome the many obstacles thrown at him while keeping his composure and thriving along the way was exemplary, and he was an inspiration to many of us. We wouldn't have been the same group without him.

Serge was very quiet at first. It was hard to break through his introverted shell. But as I got to know him better, we warmed up to each other and he confided in me that he was part of a traveling Russian circus and invited me to one of its shows. When I sat down in the bleachers and the show started, I quickly identified Serge in the opening juggling act. Next thing I knew, he was flying high on a trapeze and then performing some jaw-dropping acrobatics, all with glasses that somehow stayed put on his face. I was so excited to meet up with him after the show and gushed about his perfect hand-eye coordination, but he quieted me down. "Could you keep this between us?" he asked, not wanting the whole school to know about this just yet. "Of course," I replied, and I kept my word.

The greatest gift my freshman year at NEST+m gave me, aside from stellar academics, was a social circle. Some kids had money, but that wasn't the case for most, so no one felt left out. We were a competitive bunch, and that pushed us to excel. After school, my group of friends—which now included Serge; Wei; Lenin, from Honduras; and Jennifer, from Korea—would meet up at the corner of Columbia and East Houston Streets before heading home. In the vestiges of warmth before the impending fall, we would sporadically occupy the handball courts adjoining campus in a frenzy of excited shouts, or we'd roam the Lower East Side, exploring in adolescent wanderlust. I had finally found my people in New York.

Although she probably didn't have much of a choice, Mami was fine with me taking public transportation, which meant I was gripped by a freedom I had never felt before. My mom was working two jobs—as a home attendant for the elderly and cleaning offices with my abuelo. When I arrived home early from school, she'd ask me to start chopping some vegetables and prep the kitchen for dinner, which I enjoyed doing because I liked cooking. But most of all, I liked lending her a hand when I could. A few times a month, on those days when she was overworked and exhausted and knew I didn't have any important extracurricular activity going on, she'd give me a call and ask me to meet her at one of the offices in the Flatiron area to help her finish out that day's cleaning shift. Knowing the enormous effort she made for us, I dropped what I was doing and headed straight over. As much as I appreciated her hard work, when I left my friends to join my mom, I usually excused myself by saying, "See ya later. I have to go help my mom at the office," deliberately giving them just enough info so as not to incite any follow-up questions. Sometimes that didn't work, and someone would pipe up and ask more about her job. I'd answer, "She works in an office for some doctors," never specifying that she was there to clean. I wish I could say the contrary, but I felt a sense of shame about her jobs, mostly because that's how she felt about them herself. My sentiment was a reflection of her own frustrations. After all, she was a college graduate who had been a teacher and a principal, influencing the lives of hundreds of students, and now she was cleaning offices again and taking care of old people.

It took me years to recognize Mami's hustle. She swallowed her pride and dealt with her shame just to give me a better shot at life. Perception, or, as we call it, el qué dirán, is an enormous cultural burden and barrier in Hispanic communities. It keeps us "in line" with preconceived beliefs of restraint and fear of the unknown, and can prevent us from finding our own unique voice and using it. Only time and experience have allowed me to unlearn these beliefs and take control of my narrative, staying focused on my goals rather than on how others perceive me, which is beyond my control. In retrospect, I wish I would've had the wisdom then to tell Mami that there was absolutely nothing to be ashamed of, but I was too young to understand this at the time.

When I reached the offices, I'd greet her with a kiss on the cheek, slip on my headphones, and listen to music while we reset the workstations, dusted the surfaces, swept, mopped, and vacuumed. Sometimes I'd take a lap around the office with a Mr. Clean Magic Eraser and wipe all the scuffs and smudges off the walls. Once we were done, we'd usually grab a slice of pizza or stop by McDonald's to get an order of Mami's favorite fries. Other times we'd even gleefully splurge, grabbing a burger at the Shake Shack in Madison Square Park. With our busy schedules, this was also a chance for us to spend some quality time together. To this day, we still keep it simple, opting for a slice or some fries while we catch up on work, life, and make fun of my tíos.

My before- and after-school time was not defined by events but by a person: my chemistry teacher, Dr. Vincent Pereira, who in his humility and desire to connect with his students insisted we call him by his first name. Vincent immediately identified my group of friends in his class as overachievers, and he had a eureka moment when he crossed paths with an NYU professor, David Grier, who would swing by NEST+m to drop off his child every morning. Vincent mentioned he had just received a grant and wanted to use it to further his students' research experience, and Professor Grier agreed to help. That's how the collaboration between the NYU Department of Physics and Vincent was born.

Vincent carefully chose a group of ten students who excelled at math and were interested in computer programming. Serge—who was already in Calculus 2 with the seniors—and I formed part of the team. When the idea was presented to us, we were more than game. We were the kind of students who saw coming to school an hour earlier every day as a privilege, not a sacrifice. But that didn't mean I eagerly jumped out of bed each morning. Instead, I slowly got up and groggily lugged my inflatable mattress upstairs to store away for the day. I made a pit stop in the bathroom to clean up, and then grabbed the sandwich and cafecito Mami had prepped for me before she left for work as I hurried out the door. Then, like most teenagers, I basically sleepwalked through my hour and a half commute.

The special project's objective was to quantify and characterize the material concentrated around a specific black hole. The bright spot seen in the center of the IR (infrared) image is what we assume to be the black hole and its surrounding accretion disk.

A black hole is a place in the universe with such an enormous gravity pull that nothing can escape it, not even light, because matter has concentrated so much in this tiny area. It can happen as a result of a dying star, although theories suggest that many black holes have existed since the beginning of the universe. Their size can vary from that of an atom to supermassive, millions of times larger than our own sun, such as Sagittarius A*. Black holes cannot be directly observed, as light cannot escape them to travel to our eyes or an observatory's instruments, so they are analyzed by studying the surrounding stars and objects, which have distinct orbits and patterns when they travel nearby. The center of a black hole is called a "singularity," and the area around this will be encapsulated by an "event horizon" (the boundary marking the limits of a black hole), at which point, if an object gets too close, it undergoes "spaghettification" (vertical stretching and horizontal compression) and is absorbed by the black hole. These massive celestial bodies are still intriguing to this day. It's hard to fathom such large objects from our relatively small planet's perspective. NASA didn't actually capture an image of a black hole until 2020, when Dr. Katie Bouman and her peers developed an algorithm to make this happen with a composition of images collected from telescopes around the world.

We use infrared imaging techniques because the visible spectrum becomes difficult to detect at far distances, and most black holes are extraordinarily far from us—the closest one is a thousand light-years away. Using near-infrared techniques allows us to see objects while the dust surrounding them may remain transparent, but moving closer to far-infrared spectrum imaging allows us to see more of the dust—this is where our student research came in. We had to quantify that dust.

Every morning for an hour before first period started, we were given images taken from the Spitzer Space Telescope—a spacecraft designed for infrared astronomy that was launched in 2003, with an expected lifetime of five years but that didn't retire until January 2020—and datasets from the NYU Department of Physics to ultimately confirm if the professor's model was accurate when compared to the amount of material we observed. So we'd note the brightness of the light surrounding the black hole and use specific measurements, parameters, and math to quantify how much material was around it.

Once a month, we met with professor Grier at NYU to review advanced numerical models, which included high-level math that not even Serge was familiar with yet, as well as additional background information on the accretion disk around this black hole and why some black holes may have certain materials and others don't. Stepping onto NYU's sprawling campus and into one of the professor's labs was like setting foot in Dexter's laboratory. There were whiteboards with equations, students visualizing their data on their computer screens, posters of the cosmos, and a variety of astronomical models decorating the rooms. Then having the opportunity to talk with professors and grad students, not just about their work but also about what steps they had taken to get to where they were, was incredibly beneficial for us high school students. It was as if these grad students unknowingly became mentors for us, which had a palpable impact on me, because growing up in the Hispanic community, I had learned not to ask for help beyond my family. I had been taught to keep my head down, do my work, and not bother anyone. But I was quickly learning how insightful and necessary these kinds of conversations and mentorships were. They provided a road map on my journey to becoming an engineer.

We worked as a team into our junior year, collecting data, creating graphics, and supplying results to our findings. We had software that processed the telescope images, then we'd manually look for the specific images we needed, fill in spreadsheets with the pertaining info, and analyze the data using MATLAB code—a programming language typically used for number crunching and visualization across a variety of fields—that would later give us the statistical information we were looking for. That's basically what we did day in and day out, under Vincent's supervision. Once we had our results, we'd send our numbers to the NYU professor to get them verified. By the fall semester of my junior year, we were able to fit all our data into a model and conclude, "Here's how much material we estimate exists along the accretion disk of that particular black hole," which the professor used to validate that his general model also worked with this subset of data. It may sound a little tedious to some, but learning MATLAB and seeing advanced mathematical equations put to use in the real world was absolutely thrilling to me. After all, this was my first formal interaction with research, astrophysics, and a project related to space, and I couldn't get enough of it.

CHAPTER 8

FINAL COUNTDOWN

Education, the Silver Bullet to Poverty

There's only one optimal trajectory to Mars, but to get there our systems must withstand extreme and oftentimes unknown environments. Space is incredibly vast and unpredictable. To help our systems survive the journey, we collect data to characterize mathematical models and use them to plan around worst-case scenarios involving temperature, radiation, gravity wells, and, in the case of the rovers on Mars, topography and different types of soils. We attempt to foresee all possible issues for entry, descent, and landing to be as smooth as possible. NASA has had many successes landing on Mars, but we can't forget that there have also been some failures. Prior to the Mars Exploration Program's rovers of the early 2000s, we were shadowed by a string of several failed Mars mission attempts in the 1990s by Russia, Japan, the European Space Agency, and the United Kingdom. NASA itself saw the *Mars Climate Orbiter* burn up in the Martian atmosphere upon orbital insertion and the *Mars Polar Lander* land a bit harder than expected and suffer its own rapid disassembly event. These missions still offered learning lessons and expanded technologies that directly fed successful follow-up missions. It is important to reflect on failed missions, learn from them, and avoid making the same mistakes in the future.

When we finally concluded our black hole project, Serge and I were tapped by Vincent to accompany him to Washington, DC, for the 215th meeting of the American Astronomical Society, where we would present our findings. I think he chose the two of us because we were known in the group as the two math and science guys, the brainiacs. To take part in this conference, we had to either present a talk alongside submitting a paper and poster or simply generate a poster with our findings and conclusions. Inexperienced and frankly daunted, we opted for the latter. In the meantime, Vincent managed to secure a grant from none other than NASA to cover our round-trip train rides, hotel rooms, and meals for the three days we would be in attendance.

When we arrived in DC and surveyed the hotel's conference floor, Serge and I were blown away. Buzzing around this sweeping space with high, grand ceilings and tables in a maze-like formation were prestigious academics, scientists, astrophysicists, representatives from renowned institutions from around the world, and us, likely the only high school students in the room, just a couple of pelagatos (as Ecuadorians would say for scrubs or newbies). Even in our wonderment and admiration, I don't think we completely grasped the enormity of the moment. Our experience was nil compared to those roaming the floor; nonetheless, we were there. Vincent took us to some evening talks about how galaxies are formed and why certain stars are the way they are. Unsurprisingly, they were way beyond my comprehension, but I still tried my best to absorb what I heard.

We woke up bright and early the next day to prep our exhibit. After breakfast, we went to our designated area, set up our poster on our table, and toured the expansive room to glimpse what other attendees were presenting. These are the types of events where scientists meet up, share their results, and start collaborations through networking, resulting in projects that will likely expand our view of the enormity of the universe. While we glanced at the many posters and booths dotting the floor and skimmed data we knew was important even though we couldn't quite grasp it yet, we kept mumbling to each other, "This could be us one day!" Regardless of what field we ended up choosing, Serge and I knew we wanted to be in this type of setting, sharing ideas with colleagues to get future projects off the ground.

I had no qualms about approaching other attendees and introducing myself. "Hey, I'm a high school student. Tell me about your poster." Frankly, I led with being a high school student so they knew they'd have to simplify their explanations for me, which put some of them on the spot. Even teenage me could tell they struggled to express their research in layman's terms, a challenge many of us in STEM fields face to this day. True mastery of a subject is best portrayed when a person can stand up and explain what they're doing in simple language so the people who have no clue can actually understand it.

Serge and I took turns presenting our own data throughout the day, and we found many of these geniuses came up to us unaware that we were high school students. And even when they found out, they were still impressed that we were able to explain our findings so clearly. These experts, people we looked up to in awe, were kind, attentive, and welcoming. Though we initially had felt out of place, we began to feel like we might belong here, that we could belong here.

As that day progressed, Serge, who's not the type to complain, turned to me and said, "I don't feel well." He was visibly uncomfortable, clutching his stomach, and I knew something serious was up. Maybe it was something he ate? Unsure of what to do next, we called Vincent over, and he decided to take Serge to the hospital to get him checked out. Meanwhile, I wrapped up our presentation and headed outside for a stroll to check out the National Mall, see the US Capitol and some other official-looking buildings from a distance as the sun receded and dusk welcomed the end of that wintry day. When I entered the hotel room I shared with Serge, there was still no sign of him. They arrived a couple of hours later—the doctor had said it might be gas or a stomach bug, but Serge showed up only looking worse.

The next morning, waking up to more blistering pain, Serge decided to call his parents, who immediately got in their car, drove down from New York, and took him home. Vincent and I returned to the conference, but it was not as fun without Serge—worry was the ruling sentiment of the day. On our way back to New York, the following day, I got the call from Serge. His parents had driven him straight to the ER, where he was diagnosed with appendicitis. With no time to waste, they prepped him for surgery and rolled him away. Thankfully, he lived to see another day. Our first scientific conference was dubbed: "The Weekend Serge Almost Died."

The entire black hole project experience made me realize that data collection and research, though important, hadn't piqued my interest as much as the idea of building the tools, the spacecraft, and the systems that would enable said research. I walked away from that conference discovering other top-rated schools beyond the likes of MIT, such as Georgia Tech and the University of Michigan, and confirming I definitely wanted to be a part of the space world, but I was still unsure of what that would look like for me.

A month or so later, Mami had managed to save up enough money for us to move out of my uncle's house in Mill Basin and find a place of our own again. I had spent the last two and a half years doing my homework on the couch and sleeping in the living room, and Mami knew, as a growing teenager, I desperately needed a little more space. We had to stay within our budget and be within walking distance of a train station, so we could skip at least one leg of the commute into the city every day. After checking out a few places, we landed in a studio in Bay Ridge—a third-floor apartment steps away from the Verrazzano-Narrows Bridge, close to the R train stop on Fourth Avenue, and only a few blocks away from our old stomping grounds, where we had first landed back in 1997.

By the end of my junior year, my SAT scores were not perfect, but thanks to weekly tutoring sessions at school, they were still above average. To this day, I oppose standardized testing, and not just because I couldn't master them. These exams were never able to measure my thought process, let alone my creative problem-solving skills. Memorizing for the sake of passing an exam is an absolute waste of time. And I'm not alone. Many students who are brilliant in their own right may not excel at these types of exams because they process information differently. Circumstances come into play too. If you come from a low-income household, your focus may at times be required elsewhere; you may need to get a part-time job to help make ends meet. Furthermore, access to tutoring, practice tests, and time is what leads to success in standardized testing—all of which require resources that many students may not have. Add that to the peer pressure students face to fit in and just the general angst of being a teenager and you've got a perfect storm ready to wreak havoc on the "standards" students are supposed to achieve. This means brilliant minds may be nonsensically weeded out of a pack of top university applicants without anyone ever considering the person behind those scores. Yet another automated system that needs to be humanized.

Early on, I recognized that these types of tests weren't my strong suit, but instead of half-assing it or throwing in the towel altogether, I pivoted my strategy and began to prioritize what I needed and wanted to learn first, especially as we started entering AP exam territory. I tend to suffer from test anxiety, a symptom of performance anxiety. Even though math came easy to me, tiny mistakes, like forgetting to flip signs, have lowered my scores throughout my academic years. I do see the value in the AP content because it gives high school students a chance to take college-level courses, but those tests take away the joy of learning the material and put a ton of pressure on students. When I knew I had an upcoming light school day during the week, I'd tell my mom, "I don't want to go to school today. I need to study for an upcoming exam." And she'd reply, "Okay, mijo." I cut school to study and take practice AP tests.

My AP scores turned out fine but not outstanding. Top universities require a 5 score to get college credit for AP classes. I got 3s and 4s. Receiving those credits would've allowed me to take other classes in college or spend more time building projects, but I just had to deal with the cards I was dealt. I still feel I could've done better if I'd practiced even more, but I also now realize I was still fig-uring out how to best prioritize my time. Could I have read my text-books during my commute to and from school? Of course, but I also wanted to please my high school girlfriend, play Angry Birds, and hang out with my buddies. Fortunately, my friends were ambitious too, and I thrive in competitive environments, so that helped push me back to the books enough to get decent scores. Still, it was my first humbling lesson in how to be okay with being "good enough" rather than excellent at everything. I still strive for excellence, but now I know that the end goal does not require perfection.

As the summer break before senior year approached, someone mentioned that I should look into Cooper Union's summer STEM program for high school students. I had never heard of these types of experiences, and when I investigated it further, I discovered that several top universities, such as MIT, Virginia Tech, and the Uni-versity of Michigan, offered similar programs, and some went even as far as taking care of room and board. The deadlines for them had come and gone, but Cooper Union was still an option, so I jumped at the chance and applied, along with Serge, and we both got in! The program was completely free—all I had to do was figure out my food situation, but that's what Two Bros. dollar pizza slices were for!

Sixteen high school students from around the city were in the Robotics course. We were divided into four teams—Serge and I stuck together, of course—and were tasked with designing and building a robot whose objective would be to get the other robots outside of a four-by-four-foot ring while not driving itself out of the ring in the process. The assignment was twofold: our robot had to have a sensor suite that detected the edge of the ring and turned it around so that it remained within bounds, and it had to have a mechanism to get the other robots out of the ring.

I was more inclined to the design, manufacturing, and assembly part of the project, so one of the students in our group, Darien (who later became an industrial designer), and I took the lead on the design of the robot itself, while Serge and the other student focused on the software-programming side of things. It was a match made in robotics heaven.

When we weren't working on our robot, we took classes that taught us how to use CAD (computer-aided design) software; explored basic electrical engineering concepts, like circuit design, as well as bits and bytes mathematics and basic computational theory; and exposed us to the idea of manufacturability. You can bust your behind designing something extraordinary, but if you can't break it down into easily manufacturable pieces, no matter how cool it looks onscreen, it will be useless in the real world. Our team of instructors included a robotics professor, Brian—a short, very animated, stereotypical-looking engineer who paced across the lab while imparting his knowledge—and a few university students who not only taught class but also shared their college experiences with us, introducing us to the ideas of paid internships and undergraduate research opportunities across the country.

It took weeks and a lot of iteration to nail our robot, which immersed me in the design process and the patience required to get it right. The four of us would stay after class, racking our brains to manufacture the robot we had envisioned—time became irrelevant until the day of the competition.

When we eagerly walked into the lab where we had built our electronics for this project, we were thrilled to find a four-by-four-foot arena on the floor in the middle of the room. Students and instructors excitedly found a spot around the arena, and a person from each group kneeled on the floor and placed their respective robot in one of the four corners of the makeshift ring. We had all programmed our robots to start running autonomously with a press of a button, so at the end of the opening countdown we simultaneously set off our robots and stood up to watch with the rest of the class. One by one, each team's strategy was put to the test. Most were built to simply push the others while trying to stay within the bounds of the ring. The first one to go drove right off the border and was immediately disqualified. Another was pushed out by our opponent as ours picked up its wheels—an accidental team-up move. And then there were two. Our robot was simple and to the point: we had kept it close to the ground and had added a ramp underneath it so we could take an opponent robot's wheels off the pad and push it out of the ring. And the ramp worked! We were the last robot standing, claiming the first-place prize. That day I went from the kid who dreamed of being Dexter to the teenager who yearned to be Tony Stark.

That summer I also began drafting my personal essays and statement of purpose and even started applying for some scholarships. I was eager to gain traction in the college-application process before senior year, and I already had some top schools in mind: MIT, the University of Michigan, Georgia Tech, Cornell. I wanted to stay on the East Coast to be closer to my family. Meanwhile, my mom started to familiarize herself with the financial-aid system because we could not afford to pay the fifty thousand dollars a year each top-notch school charged. We had to find another way. I accompanied Mami to a financial-aid workshop she had found that was taught in Spanish inside a church in Brooklyn. This helped us both gain a clearer understanding of the process, and although she couldn't review my essays because they were in English, she'd check in on me periodically to make sure I was hitting the necessary deadlines. It was a team effort, and as always, she encouraged me every step of the way.

Before heading into senior year and then off to college, I was dead set on finally making my way back to Puerto Rico for the first time since we'd left more than three years earlier. I had started tutoring a few NEST+m middle schoolers, which allowed me to save up enough to buy myself a kick scooter (my friends relished making fun of my new mode of transportation) and a round-trip ticket to Puerto Rico in August.

Sonia picked me up at the airport, and it immediately felt like I had never left. I visited Niní, played video games with Benito, and watched documentaries with Robert. Since classes in Puerto Rico start in August, I spoke to the principal of Notre Dame, the same man who had been there when I attended, and asked if he'd allow me to come to school for that month. It was the only way I would be able to spend a good amount of time with my old friends. He considered it and said, "Yes, as long as you wear appropriate attire to match their uniforms." That same afternoon, I went to Marshalls and bought a couple of shirts and pants, and the following day, Sonia drove me to school, where I began senior year with my crew from Caguas. I attended Chemistry, English, Science, hanging out like old times. At lunch, I sat outside with my friends and talked about the schools we were planning to apply to that fall, and I breathed a quiet sigh of relief when I realized that despite being apart for the last four years, we were still all thinking along the same lines. It confirmed that I had managed to regain some of the future that I thought I had lost when we moved back to New York, the future that included my friends from Puerto Rico.

On February 28, 2011, I came home from school, opened our mail-box, and found a thick envelope stuffed inside together with some junk mail. My spine tingled as I quickly pulled it out and searched for the return address: the University of Michigan. I went up to our apartment, opened the envelope, and scanned the acceptance letter. My first response from a university was a positive one! While I was pleased, I honestly didn't think much of it because my dream school was MIT. So I set aside the package on our kitchen table and shared the news with Mami when she arrived from work. Nothing could wipe the look of pride off her face. One down, several more to go. Every morning, prior to class, my friends and I got together to share our latest college news and celebrate someone's boom while comforting another person's bust. By the time I heard from MIT, it was in the form of the thin envelope we had all begun to dread. Rejected. *How will I become Tony Stark if I can't go to MIT?!*

In the midst of my devastation, letters from Ivy League schools followed. Columbia, rejected. Princeton, wait-listed. Cornell, rejected. I was crushed. But I had to remind myself that I still had options, and pretty good ones no less. (To be honest, it also helped ease my pain when even our class's valedictorian didn't get into MIT.)

I narrowed down my choices, based on financial aid, to the University of Michigan, the University of Virginia (which had offered me a presidential scholarship!), Georgia Institute of Technology, and Carnegie Mellon. When I read that Carnegie Mellon was willing to fly out prospective students to show them around the campus, I packed my weekend bag and headed to Pittsburgh. During this student tour, I met professors and alumni, and stayed with current students in their dorms. I also met some Puerto Ricans and other kids from New York who were going to Carnegie Mellon, which was supercool—I found great comfort in coming across other students who looked and sounded like me. But when I walked into their Electrical and Computer Engineering department, I was swept off my feet. First, to get there, we had to cross the Pausch Bridge, aka the Rainbow Bridge, which features more than seven thousand programmable LED lights along the walkway that illuminate the entirety of the structure with a looped multicolor light show—I felt like I was strolling through the middle of a rainbow. Then we entered the impressive Gates and Hillman Centers with Guggenheim-like spiraled indoor pathways that led to fully equipped labs and classrooms. That's when I pretty much asked where to sign on the dotted line. I accepted their offer that weekend and began to lean toward majoring in electrical engineering, since they have one of the top programs in the country.

Back in Bay Ridge, a few days after my eighteenth birthday, I received an unexpected call. It was Darryl Koch, an adviser at the University of Michigan. "We haven't heard from you about this M-STEM program you got accepted to. Are you going to come to Michigan?"

"I have no idea what you're talking about," I replied. Then it hit me: when I applied to Michigan, I also had applied to a separate summer program I had found interesting. I could go to Michigan for six weeks during the summer before school started to get to know the school, the area, and the dorms and get acquainted with other freshmen. "I never heard back, so I assumed I didn't get in."

After a few minutes of Darryl clicking on his computer keys to verify my information, we realized they had the wrong address on file.

"This all sounds great, but I already accepted Carnegie Mellon's offer. I never had the opportunity to fly to the University of Michigan to check it out, and I have no way of making it out there now in person."

"Well, what are you doing this weekend?" replied Darryl.

"Uh, I'll be here in New York," I said hesitantly.

"Okay, would you be interested in flying out to Michigan?"

My jaw dropped. "What about Carnegie Mellon?"

"It doesn't really matter. We're going to fly you out to Michigan so you can see what you think for yourself."

"Um, sure. Sounds good!"

Darryl himself picked me up at the airport, drove me to campus, and walked me around, showing me all the points of interest. He was a soft-spoken yet firm man, who was also attentive and warm. His engaging personality and key guidance on how to negotiate the offers on the table to better serve me made a huge difference. We later met up with two prior M-STEM students who, together with Professor Brent Gillespie, were part of the Haptix Lab, based in the Mechanical Engineering department. They enthusiastically showed me around the lab and the current projects they were developing, while sharing their experiences, and I was like a super-absorbent sponge, soaking it all up. More than anything else, what drew me in was the human connection. Darryl then walked me to where some of my classes would take place as well as the M-STEM dorm where I would stay that summer, should I accept this offer.

But what about the cost? Finances, or lack thereof, had always been the most important piece of the puzzle. "We'll work with you to increase your financial aid," he said with such certainty, and as a wave of relief washed over me, I thought, *How can I say no to Michigan now?* M-STEM, their program designed to support incoming students in their transition from high school to college and during the first two years in school, plus the overall kindness and friendly vibes, were the real clinchers.

When I got back to New York, I said goodbye to Carnegie Mellon and hello to the University of Michigan. A new path lay before me, and like most seniors, my mind remained in my not-so-distant future, which made it super difficult to do homework, study for the remaining AP exams, finish out the school year, and graduate. But despite the best efforts of senioritis, my friends and I managed to do it.

Our graduation took place, of all places, at Cooper Union, a few weeks before our final Regents Exams. We walked onto the stage, received our diplomas, took pictures, and goofed around in our gowns. Our tight-knit group was brimming with excitement about our future but also heavy-hearted, knowing we were saying goodbye to being together in the trenches of school day in and day out. At least we still had college breaks to look forward to.

Once the ceremony was over, a close group of us filed out of Cooper Union and headed out to eat with our families. We chose Dallas BBQ and had a big ol' meal in celebration of the end of this season in our lives. As we dug into our barbecue specials and the mountainous scoops of coleslaw and mac and cheese on our plates, the idea of leaving home didn't fill me with dread or melancholy. On the contrary, I couldn't wait to get started on this new chapter in my life, the one I had been dreaming of since Puerto Rico. When the check came, my grandfather, who was there with my grandmother, quickly grabbed it from the waiter and said, "I'll take care of it." Serge's parents insisted on splitting the total, but my grandfather wasn't having it. I had never experienced a gesture on that scale from him, and I knew that paying the large bill was an even bigger sacrifice on his part.

In many ways, it was a reflection of all the sacrifices my family had made for me to reach that moment in my life. Though I had done my schoolwork on my own, my journey had always been a team effort. Securing my future was the way I could pay them all back, especially my mom. It was my chance to let her rest while I began to work. And now that moment—*our* future—was finally within reach.

CHAPTER 9

LIFTOFF

Skyrocketing to My Dreams

From the moment I stepped foot in New York until college, I never had a room to call my own. I was always working toward my future, the way Mami taught me, and I was somewhat reliant on others along the way, which at times made me feel like a burden, especially to Mami, who had to juggle so much while raising me. I felt indebted to her for the sheer sacrifice she had endured to give me a better shot.

Once I was gone, Mami began to reconsider her circumstances in New York. She was on the verge of turning sixty, and those long days taking care of elders and then cleaning offices in the evenings were catching up with her. She'd come back to the now empty apartment late at night absolutely beat, only to have to set the alarm and do it all over again the next day. Each morning became harder to endure. Simply standing up from her bed or a chair would suddenly provoke a back cramp that drained the color from her face. Her body was screaming, "Mayday!" and this time she listened. At a doctor's appointment, after explaining her symptoms, he gave it to her straight: "Your body is telling you it's time to stop." That statement kept running through her mind during the long subway ride home after clocking out of yet another evening work shift. Trepidation hit when she reached her stop. Those late-night walks from the station to her building caused her anxiety to spike. After mulling it over and talking with Xavier, who insisted she move in with them, she decided the time had come for her to hang up her boots and apply for early retirement. That also meant saying goodbye to her Bay Ridge apartment and leaving New York one more time, this time for Florida.

We continued to speak daily and remained as connected as ever. It took time for me to fully understand and accept that now I was completely independent financially and living on my own; what happened next was exclusively in my hands. The classic American dream suddenly felt just within my grasp, and I wasn't about to squander it.

When I look to the stars of engineering, my personal constellation, I see Clarence Leonard "Kelly" Johnson, who graduated from the University of Michigan with a master's in aeronautical engineering in 1933. A legendary aeronautical and systems engineer who worked on Lockheed's U-2 and SR-71 Blackbird, his story was already familiar to me by the time I arrived at Michigan. I had nursed a boyhood fascination with planes, enamored by their history and the scientific strides made possible by their innovation. Aerospace Engineering, the university's internationally renowned program, encompasses both aeronautics and astronautics—in other words, it's a discipline that includes the design of both aircraft and spacecraft. From a historical standpoint, aircraft came before spacecraft.

From a technical standpoint, the former are designed to travel in the atmosphere before reaching space, while the latter need to be able to fly in low Earth orbit or beyond. Furthermore, each vehicle has different environmental constraints as well as boundary and operating conditions that must be considered and reconsidered to get these pieces of machinery off the ground and to their destination.

An engineer uses creativity, bounded by known and understood laws of nature, to solve simple and complex problems. We may not all be as cool as Tony Stark in *Iron Man*, but our brains work at a similar level. When engineers see something, we're trained to think, *This is good, but it could be better*. Ours is a language of imagination and construction—we visualize a system, break it down into smaller components, distill the internal workings, and problem-solve our way to completion.

The M-STEM summer program at U-M, as a precursor to college, gave me a glimpse of the rigor to come by providing a deep-dive simulation of the pressure I would feel while taking four engineering classes at once (Physics, Math, Programming, and Engineering Projects). But M-STEM connected me to the necessary resources to succeed, from organizations to advisers, and even stipends to support lower-income students.

One of the organizations I took note of during M-STEM was the Society of Hispanic Professional Engineers (SHPE), when I met the local chapter's president, Paul Arias, a tall Peruvian with curly hair and a contagious laughter who was a mechanical engineer on a path to his PhD. He was always quick to help others on a professional or personal level and exuded a warmth and charisma that made everyone feel at ease in his presence. Paul explained that when we began the fall semester, SHPE would be there, ready to help us every step of the way. And they were.

I was first drawn to SHPE for the professional development and mentorship it offered, but soon the group became my familia. We were there to push for greater inclusion and elevate one another in the process. Promoting outreach to young Hispanic students, SHPE organized its annual Noche de Ciencias at a nearby high school in Detroit, which had a high Hispanic population among its students. We tried to stoke interest in the event by visiting twice beforehand and talking to students about engineering, but of the forty guests we had estimated would attend, only twelve showed up.

Hispanics are resilient, hardworking, and hopeful. It's engrained in our culture to put family first and care for one another. Oftentimes, however, many of us face challenges based on simply being Hispanic. Note the following key statistics I came upon while doing research for a talk in a SHPE conference in 2022:[*]

- During the 2015–2016 academic year, 44 percent of Latino higher-education students were the first in their family to attend college.
- In 2021, 19 percent of Hispanic children under the age of eighteen lived in households where no parent had completed high school.

---

[*] Postsecondary National Policy Institute, "Factsheets: Latino Students," September 19, 2022, https://pnpi.org/latino-students.

- Of Latino higher-education students in the 2015–2016 academic year, 76 percent received some type of federal aid, compared to 67 percent of white students, 50 percent of Asian students, and 88 percent of Black students.
- Of Latino higher-education students, again in the 2015–2016 academic year, 60 percent received a Pell Grant. Overall, they made up 22 percent of all full-time Pell Grant recipients.
- Of Latino students enrolled in higher education in 2018, 33 percent worked forty hours or more, 19 percent worked thirty to thirty-nine hours, 26 percent worked twenty to twenty-nine hours, and 21 percent worked one to nineteen hours.

These statistics tell the stories of many of us who are first-generation college graduates in the United States, navigating education in an expensive system that heavily relies on networking through activities that are often also financially out of our reach. Many students don't even have mentors, or know to look for them—a crucial step since many of us don't have parents or family with the experience to help us succeed in these areas. To make ends meet, students may take on one or more jobs, which ultimately makes keeping up with academics an unattainable challenge. Many students end up as survivors, not thrivers. It's best captured by one of my mentors, longtime friend and now a professor at Brown University, Dr. Mauro Rodriguez.

Per Mauro's research and data from the American Society for Engineering Education (ASEE) and the US Bureau of Labor Statistics, there are approximately 87,000 Hispanic undergraduate engineering students. Approximately 50 percent of those students end up in an academic burnout cycle, accompanied by financial struggles due to inherent lack of accessible scholarships and resources because of lower GPAs. This, in turn, gives way to inaccessible professional and research opportunities, ultimately leading many to fully remove themselves from their STEM majors. About 18,000 or so graduate each year, and fewer than 7,000 enter the engineering workforce. Fewer than 6,000 Hispanic students go on to obtain a graduate degree every year. Without graduate degrees there are limited opportunities for upward mobility. This is reflected in the overall number of Hispanic executives, which amount to less than 2 percent of all US executives. As I often mention in the college success workshops I lead, these issues have solutions. There are groups, programs, and organizations across the country, such as M-STEM in my case, that provide academic safety through workshops on study methods and best practices in STEM fields, as well as tutoring, supplemental instructions, and mentors. These resources are crucial in teaching students—especially those who are the first in their family to go to college—how to navigate academic culture while also maximizing access to student financial aid, services, internships, and other worthwhile engineering experiences tailored to their individual needs so as to drive their success forward. The 2020 Census data suggests that Hispanics make up 19 percent of the total US population, yet we represent only 8 percent of the STEM workforce. STEM careers have generated wealth across the United States and will continue to do so as technology continues to explode in development. For the Hispanic community to be proportionally represented in the STEM workforce, equity must be established first. It's not just this data that motivates me to continue doing outreach and inspiring young folks to pursue careers in STEM; it's knowing that one event, one talk, one call has the possibility of changing a person's life forever.

At the end of my first Noche de Ciencias night, after building towers out of popsicle sticks within a time limit to see which one would be the tallest and structures that would protect eggs from breaking after they were dropped from a certain height, all while my SHPE peers and I filled in the twelve students who did show up on the different STEM career paths available to them, a girl came up to us, tears in her eyes, to say thank you. She explained that at home, anytime she brought up her dream of becoming a doctor, her immigrant parents shot it down, saying college was way beyond their means and focusing on getting a stable job after high school would likely make more sense given their circumstances. There are workshops for parents in English and Spanish—we'd had one at this event—that teach them about requirements for college admission and financial aid, including deadlines and realistic expectations, much like the one Mami and I went to in that Brooklyn church. Immigrant parents often work endless hours and have no time or energy to consider the possibilities beyond what they already know, but that doesn't mean those possibilities aren't out there. This was the first time this girl had someone tell her that she could be an engineer, a doctor, or a scientist, if she set her mind to it, and that she could find financial aid packages that would allow her to do so. She was overcome with emotion when she realized the possibility could be within her reach after all, and it made us realize that even if only one student had shown up that night, that meant there was one person's life we had the chance to change. Two years later, that high school student enrolled at the University of Michigan.

In my very first Aerospace Design class, students were tasked to design a blimp that would work on Mars and then scale it to Earth's environment so we could test it. This was my first exposure to a full-blown design-build-test flight project. The students were divided into groups of four, and I lucked out with a formidable team: Kevin, who now works as an aerospace engineer at Northrop Grumman; Jerry, who now has a job at SpaceX; and James, who went on to work for several aerospace consulting firms.

None of us knew each other at the time, but we easily fell into a rhythm working together. James took the lead in programming the controls needed to navigate the blimp. I took the lead in designing the actual structure of the blimp and globe, and then manufacturing it. Jerry helped calculate the volume and amount of helium we would need to sustain the specific shape we were working with. And Kevin led the division of tasks to complete the design report, keeping us organized along the way. I couldn't have hoped for a better-balanced team.

The competition was held toward the end of the semester, prior to the pressure of final exams, in the Aerospace Engineering building's spacious atrium. The first competition involved getting around certain obstacles in the shortest amount of time possible. At the helm of the controls, James worked his magic. He skillfully thrust the blimp forward, adhering to our calculations, and then slowed its velocity right before it reached the end of the path, so that it could turn in place. Once it hit the return position, he throttled it up again and brought it home. Subverting expectations given its massive structure, we managed to get our vehicle to the end of the atrium and back in nineteen seconds, breaking the record for the fastest blimp on campus, a record that held for the next four years.

Exhilarated, in the aftermath of our triumphant win, all three of my team members declared their major in aerospace engineering. But a design-build-test-centric curriculum felt more in tune with what I had always dreamed of doing. I had heard that as the aerospace engineering major progressed, it became increasingly theory intensive. I had been struggling with Honors Calculus 3 that semester, and the last thing I wanted to do in the next three years was plunge even further into math. Despite being keen on working in aerospace, I figured I could gather that experience through practical internships rather than a treacherous onslaught of theory-based courses. On March 14, 2012, I declared my major as mechanical engineering.

That same spring, our SHPE chapter hosted the Regional Leadership Development Conference. Paul Arias was the chair, and I applied for the volunteer position of marketing chair and got it. As a young freshman, this was my first time participating in one of SHPE's conferences with this level of responsibility, and I had no idea what I was doing. Thankfully, Paul mentored me through the moves and infused me with the confidence I needed to take the initiative. I created a sponsorship package that highlighted our needs and how a company would benefit by sponsoring our event. I was also the one who put together promotional videos and posters, and even used social media to spread the word, which was a relatively new promotional tool at the time. Paul led our meetings, and I would chime in to discuss our progress on the marketing side of things. I'd also host calls with executives of companies who had previously connected with SHPE and now wanted to sponsor us, and I organized the use of their logos on our ads as well as their individual tables at the conference's closing career fair event.

At the end of the day, I managed to bring in around $30,000 for the conference, and whatever was left over went straight to our chapter. Around three hundred students from neighboring states came to Ann Arbor that weekend, breaking the record for most students attending a Region 6 annual conference. It was such a successful turnout, and I had enjoyed the process so much, by the end of that weekend in March, I was bent on running for external vice president—a position in charge of corporate relations and the next Noche de Ciencias. A position I would win. Rising in this role not only honed my comfort with talking to executives but also gave me the chance to start connecting students with corporate partners and provide them with possibilities.

While I was busy embarking on this new college journey, Niní's health had taken a turn for the worse. Her son reached out to my mom to see if she would consider moving in with her. He explained that Niní's current home attendant had been somewhat negligent, and he didn't want his mom living under those conditions. After everything Niní had done for us, it was an absolute no-brainer for Mami. She had the experience as a home attendant herself and could easily fly to Puerto Rico from Florida, where she was currently dedicating her time to her grandchildren.

Niní's son owned a condo on the beach, so Mami moved in with her there, and he covered all expenses and food. So these two dear friends were able to spend the next year together again, forging new routines and deepening their relationship. I got to spend some precious time with both of them during a visit to my beloved island that August before the start of my sophomore year at U-M.

At U-M's renowned engineering career fair in the fall semester of my sophomore year, I struck up a conversation with a GE Aviation rep. He was Puerto Rican, and I had already been in touch with him through my position at SHPE. I mentioned that I was interested in working with defense engines, and a seed was planted. Before the end of that semester, I already had a summer internship offer from GE Aviation. Internships in this industry are practically equivalent to a full-time job, with salary and benefits and all. The concept of unpaid internships as an engineer are unheard of, so I knew I could continue to support myself during the summers without Mami's help, which was a huge relief. Mami gifted me money for some meals here and there, but I was flying solo with no safety net otherwise.

Between this offer and the beginning of my mechanical engineering curriculum, I felt steady, confident that I was heading in the right direction. I began to study the two fundamental subjects in my field: statics and dynamics. Statics covers how forces are applied in static structures. Dynamics are used to analyze the forces in systems that are in some type of motion, following Newton's laws of motion. I loved the dynamics course, from drawing and defining known variables and assumptions to trying to uncover some unknown aspect of the system in question—it only confirmed that I had chosen the right major. Everything seemed to be falling into place.

But sometimes in life moments of great reassurance are punctuated by solemnity. I was walking across North Campus on a warm spring day, with white clouds dotting the blue sky, when Mami called me. "Hijo, Niní passed away today." My heart sank to my feet. I was aware that her physical deterioration had accelerated in the last few months, rendering her incapable of driving, and I had known this call would eventually come, but no one is ever fully prepared to let go of a loved one. That initial shock is inescapable. After I hung up and put away my phone, I walked toward the nearby Lurie Tower, North Campus's beacon, and sat between its columns to take a minute. I recalled Niní's generosity, her love, the good times we'd had with her, the way she had taken us under her wing, and I quietly thanked her for it all.

Niní's loss was the third big loss I had experienced in my nineteen years of life on this planet. The first one—Noemí, Ruth's mom—left an indelible mark on my mind. I don't think anyone forgets the first time they see a lifeless person they once knew. But the first time I observed and experienced the depths of grief was with the passing of my abuelita, who had left us only a couple of years earlier.

Abuelita, my great-grandmother, had migrated from Ecuador while we were living in Puerto Rico and once again had become a fixture in our lives when we were back in New York. Mami and I would visit my grandparents and her every Sunday, and although she was already in her early nineties, Abuelita would take over in the kitchen, feed us, and wander around the room mothering us while the rest of the adults sat at the dining room table catching up and playing card games. Until the fall of 2008, she was in good spirits, then her health began to decline. After several tests, she was diagnosed with pancreatic cancer. Advanced and aggressive, her sickness swept in like an unforeseen hurricane, knocking her down and leaving her bedridden within a matter of weeks. It was our turn to take care of her.

By early December, she was practically unconscious from the medicine used to lessen her pain, but she could still hear us. Knowing it was only a matter of time, we gathered in her room and took turns sharing old anecdotes, anything that would make her smile inside and feel the love that surrounded her. When it was my turn, Mami came into the room with me and grabbed my hand. I approached my dear abuelita and thanked her for everything she had done for us, for taking care of me as a baby, feeding me, and nurturing me while Mami went to work in Ecuador. "I love you and I'm going to miss you, Abuelita," I added. That evening, within the hour, she grunted loudly and took her final breath. We were all in the room when it happened, surrounding her with the love she had bestowed upon us for so many years.

My grandmother, aunts and uncles, and Mami burst into mournful tears. I'd never seen so many adults cry at once. It shook me. I just stood by with my cousins in silence, watching as they gave free rein to the emotions they had held back to show my abuelita a semblance of strength during those last few weeks of her life. My grandmother wept like a little girl—losing a mom, no matter how old you are, must be excruciating.

I can only imagine the pain my own Mami must have felt as she broke the news of Niní's passing to me. After all, Niní had become a mother figure in her life over the past decade and an all-important member of our family. Both my abuelita and Niní lived long and wonderful lives, and neither of them passed away alone—I take great comfort in that, and it allows me to not linger in grief for too long, something I didn't have the time, space, or capacity for if I wanted to stay on my trajectory. *Que en paz descansen, Abuelita and Niní.*

When I was younger, my mind cluttered with documentaries and history programs, all I could see in my future were jet engines for fighter and military aircraft. In the United States, an inextricable tie exists between the government and aeronautical innovation. From the accelerated evolution of airplanes during World War I to the Space Race—even the Apollo moon program was entirely based on showing the Soviets the United States could reach and weaponize it if need be—this was already abundantly clear to me by the time I got the summer internship at GE. Most groundbreaking jobs in aerospace are highly regulated. The innovative and military technologies developed at GE are restricted and can't be exported without proper documentation and processes, so working on them is limited to US citizens only. Without that clearance, working at these companies is still viable but at a much more limited capacity.

As a green card holder on the path to citizenship, I was hired to join the Inlet and Exhaust Systems team, which required no clearance. In a nutshell, the team works on the systems around the main engine, the powerhouse that converts fuel into thrust, to keep it cool, silent, and minimally traceable for stealth developments. This was an amazing opportunity right out the gate, and it was the first time I had the chance to apply what I had been learning in class to real-world projects, at one of the United States' top aerospace companies no less. It was all starting to click, and I couldn't have been happier.

CHAPTER 10

MISSION CONTROL CENTER

The Data That Helped Shape My Identity

Once I was finished with the internship at GE, I packed my bags, shipped some of my stuff to New York, and then boarded a flight to Ecuador, returning for the first time since Mami and I had left sixteen years earlier. As I gazed out the porthole, I wondered, *Will I even be able to recognize my dad?* He wasn't active on social media, and although we'd shared occasional phone calls, I hadn't seen him in fourteen years. Walking through the arrivals gate, I scanned the crowd and immediately locked eyes on his tall yet somewhat older and weakened figure. When I approached him, I noticed the cane in his hand—he was still recovering from a tragic car accident that had killed his brother-in-law and almost left him paraplegic. Emotions didn't really stir within me; it was weird, almost like greeting a stranger. We then gave each other a strong hug, sharing thumping heartbeats through our chests rather than words. As we parted, it hit me: I was taller than him now. By his side stood my uncle—Dad's brother—my aunt, and their four children, who were all around my age, give or take a couple of years, cousins I was meeting for the first time. "Hola," "Mucho gusto," "Nice to meet you," we said and gave one another a kiss on the cheek. There I was, back in Ecuador, walking toward the airport parking lot with my father on one side and my newfound family members on the other. My emotions were on ice, numb from the enormity of it all.

I had requested that my first meal back be bolón, an Ecuadorian version of Puerto Rico's famous plantain dish mofongo, but with gobs of butter, no garlic, and topped with queso fresco. Even though it's typically served at breakfast with a cafecito, I was ready to lap up that dish morning, noon, and night. So my family indulged me and took me to El Café de Tere, which is a chain restaurant with the most delicious bolones you can imagine. As we sat down with our food and dug in, my cousins' presence broke the ice and we hit it off right away. Everyone seemed to relax into one another's company as the minutes passed quickly. My dad started up with his signature banter, and I suddenly ascertained that I got my warmth and knack for making people feel at ease from him. Without losing a beat, I jumped into the chatter, and we kept the table talk buoyant and light. Dad seemed somewhat frail, his dark hair now streaked with flashes of gray, but his smile remained intact and unchanged.

We spent the next few days exploring in and around the city together with my cousins and anyone else in the family who wanted to join us. Certain blocks and stores felt incredibly familiar, and others were completely new to me. But then we turned a corner and parked across the street from what had been my first home. The house looked like a smaller version of what had been etched in my mind, strange but still recognizable. I would've liked to go inside, but Mami had been a renter back when we lived there, and we didn't know the current owners. It had a relatively tall wall between the street and the house, with one door that led to a tiny front yard followed by the house's front entrance.

As I stared at my old home's facade, I remembered the time I had decided it would be funny to hide behind the door of one of the rooms in the front of the house where I could spy on the adults from a nearby window . . . right before my dad was scheduled to pick me up and take me to school. When my abuelita called me outside to get in the car, I stayed still and extra silent, trying to make myself invisible, to remain undetectable, like a superhero. She called for me again. Nothing. Then I noticed her voice became more frantic. "El-iooooo! Where are you?" She paced the house, glancing into rooms, checking the bathroom, everywhere except behind that particular door. As the minutes ticked by and there was still no sign of me, my dad jumped out of his car and joined the frenzied search for his missing boy. With no luck, he called my mom, who was already at work, believing that their worst fear might've just come true: "He may have been kidnapped!"

Things quickly escalated from there, but rather than come out to stop their intensifying despair, I curled up into an even smaller ball near the floor, now fearing how they might react once they realized I had been there all along. Minutes felt like hours until, at long last, someone moved the door and found me hiding behind it. "He's here! He's here!"

Overcome with relief, followed by exasperation and then fury, my dad dragged me out of my hiding spot and gave me a whipping with his belt. He then deposited me on the back seat of his car, called my mom to tell her I was fine, and drove me to school in raging silence. Rubbing my behind in pain, I giggled under my breath for having pulled it off. *Got you!*

Outside of knowing Mami's side of the family, I didn't have a full picture of where I came from, so I was eager to connect the dots, meeting with several people who had known me as a toddler, including my godmother. My dad and I had lunch with her one day, and she was primed with a bunch of questions, eager to hear about what had happened since we moved to New York. From what I recall, it was probably the only time during that trip when I really dove into what it took for me to get into U-M, and what I was doing with this opportunity now that I was a full-time student. I talked about my recent internship and my path to becoming a mechanical engineer. She listened attentively, throwing out some good follow-up questions where needed, while my dad, hunched over in his seat in silence, nodded occasionally or let out a "Qué bueno," although I got the feeling that he didn't quite understand what I was talking about. But I didn't want to jump to conclusions. I was still overwhelmed by the experience as a whole, so I just went with the flow and tried to focus on seeing everything, eating everything, and becoming reacquainted with my origins. To complete this exploratory journey, we needed to make one last stop: my paternal grandmother's farmhouse in the Galapagos.

After the two-hour flight, my dad, cousins, and I drove to a small plot of land beside a mountain, and from a modest house in the midst of the jungle vegetation emerged a tiny, somewhat frail-looking elderly woman with an average build and a wooden cane in one hand to alleviate her arthritic joints. When I approached her, she cranked up her head and said, "Elito, I'm so happy to finally see you!" I bent down and we embraced in a long, warm hug. As I burrowed my face in her thick graying hair, I let out a quiet gasp—I really did get that texture of hair from her, as well as her bushy eyebrows, which I noticed when we finally released each other. There was also a clear resemblance between her and my dad in the cheekbone area. I continued to connect dots in my mind. Her spirit was joyous and loving, and she immediately treated me like part of the family.

My grandmother lived on her own in that small house in the middle of nowhere, but thankfully a family who worked the land lived nearby and gave her a hand whenever needed, and my dad's younger sister, another incredibly kind and warm woman, was only a thirty-minute drive away. So having several of her grandchildren under her roof that day made the years melt away from her demeanor and reenergized her soul. She quickly ushered us through the house, straight to the orange trees out back, pulled one from a limb, and grabbed a pocketknife, cutting a hole in the top and handing it to me. "Drink it," she said. I squeezed the orange, and drops of the sweetest juice I have ever tasted made my mouth tingle with joy. As the rest of the family settled in, she eagerly led me farther into the backyard area and pointed to a couple of enormous five-hundred-pound turtles roaming the land. When we took a step too close for their comfort, they retreated their heads into their shells, producing a hissing sound that reminded me of a truck's hydraulics system. I was completely transfixed by these dinosaur-like creatures. I was honored that she had pulled me away from the group to share these creatures, and this moment, just between the two of us. The rest of my time there was spent either at the table with everyone else or with my cousins, who took me on stunning hikes, exhilarating swims, and unforgettable boat trips to neighboring islands.

When our time together came to an end, I gave my grandmother one last hug and said, "I'll be back to visit soon." And though I did return a few years later, by then she had already taken her last breath and said goodbye to our planet. I am grateful that the one and only time we shared together was so memorable and magical.

I flew back to Guayaquil with my dad and cousins, and the following day they took me to the airport. I thanked them all for the great time I'd had and then turned to my dad and said, "Gracias. I truly appreciate everything you've done to make this trip so special. I'll see you soon. I'll come back soon." He met my eyes with a smile and replied, "De nada, Elio. We'll keep in touch." The air continued to be light and happy, so I turned and walked to the security checkpoint carrying with me a deep sense of satisfaction and joy. The exploratory mission had been a success. I had seen my father again, met new family members, and reconnected with my roots. Sure, my father and I hadn't had many opportunities to be alone during that week, but neither of us had sought them either. I was barely twenty years old and didn't have the need to go deep with him. All I really had wanted to answer was *Who am I? Where did I come from?* And that trip did just that.

While my reunion with my father didn't shift our relationship in any real way, it did open the door for conversations that would come later. On my next visit after grad school, during one of our walks around Guayaquil, we had a matter-of-fact conversation. As an immigrant for so much of my life, I often felt like I was occupying space that didn't truly belong to me, so when it came to my relationship with my dad, I just needed to understand where I fit in the puzzle that was his life. As he reluctantly relayed the information regarding his different relationships and children, I couldn't help but say, "Qué clase de cabrón eres." He laughed off my "Damn, what a bastard" hits the same way he had laughed off Mami's fear when he drove too close to the edge of the highway on our way to Riobamba over a decade ago. Although I still harbor some resentment of his lack of emotional capacity to even recognize the effect his decisions had on Mami and me, I am slowly processing that the apology I hope to get one day will likely never come. If anything, that matter-of-fact conversation gave me clarity on his story and who he is, which in turn helped me close the circle of how I came to be, and *who* I came to be, in this world.

CHAPTER 11

MAX PRESSURE

Inching Closer to Real-World Results

SpaceX will be my next adventure. Dreams do come true. I can't believe this. I am so blessed. #occupymars." That's what I posted on social media when I learned SpaceX had accepted me as a 2014 winter intern. And so began my time in Los Angeles. From a technical perspective, the SpaceX experience was a game changer. It was my first time working in the space industry, on spaceships, with none other than Mauro Prina as my manager, an Italian engineer who had previously worked at NASA's Jet Propulsion Laboratory, specifically on the Mars rovers *Spirit* and *Opportunity*. He was part of the team that had designed their thermal control system and was now the director of thermal dynamics for the *Dragon* capsule program at SpaceX. On my first day, my group met for orientation in the assembly line area of the rocket factory, the company's headquarters, and we wandered right by the rockets they were working on. I was like a kid in a LEGO store! Although SpaceX was still relatively new in 2014, I had been glued to broadcasts of their successful launches over the last few months and had taken note of the notoriety they had begun to accrue because of this. And now I was here, in the flesh, walking through a rocket factory, a gateway to the stars.

The chief of engineering at the time for the *Dragon* was none other than Elon Musk himself. The updated *Dragon 2* is used to this day to resupply the International Space Station, but the next big project at the time was to make a human certified *Dragon*—in other words, to eventually be able to transport humans, not just cargo, to the ISS. This would be a first, given that—since 2011—the United States had been relying on Russia to get astronauts to space. But the price was beginning to weigh on the United States and eventually became a reoccurring contending negotiation due to the geopolitics of the times.

Space shuttles originally were intended to be reusable, reliable, and relatively economical spacecraft, yet they are considered some of the most complex spacecraft ever flown. Unfortunately, many of the parts couldn't easily be refurbished as intended, so the Space Shuttle program became very costly over the years, marred by tragedies. Eventually President Obama decided to retire the program in 2011 and open the opportunity for private US companies to develop transport to the ISS. SpaceX and Boeing were two of the companies that vied for a piece of this pie.

We spent the next nine years without a US-based launch for our astronauts, until 2020, when SpaceX completed their *Demo-2* mission. Although Boeing hadn't been able to complete the development of its Starliner spacecraft intended for the same purposes, a Starliner is set to launch in 2023, and the company continues to have a governmental contract because both NASA and the US government are interested in diversifying their capabilities rather than relying on solely one company. So that's where we were when

I walked into my first day at SpaceX, still working on a capsule that would eventually transport humans to space.

In our Thermal Control Systems team, half the people worked on the *Dragon* program and the other half on the *Falcon 9* rocket. To get to what is now known as *Crew Dragon*, several big testing campaigns had to be completed, to demonstrate that SpaceX could develop a human-grade spacecraft. I was assigned to the *Dragon* program as a build engineer for the thermal system of the pad abort capsule. The idea was that if the *Dragon* capsule was suddenly in danger on its rocket pad—because, for example, the rocket was about to explode or another extreme scenario was about to go down—it should have the capability to launch and separate itself from the rocket to keep the astronauts out of harm's way. SpaceX had developed SuperDraco, a thruster engine to be used as part of the launch escape system in case of emergency. When I arrived on the scene, they were in the middle of building, testing, and certifying it all.

My work hours were varied, but I'd usually clock in at 8:00 a.m. and leave by 7:00 p.m. They were long days, but we had perks, like our own barista and chef, who served affordable and tasty food. There was also a gym nearby, so many of us would finish our workday, exercise until around 8:00 p.m., and then catch the last shuttle back to our apartment complex at 8:30 p.m. This was the first time I started lifting weights with my roommates' guidance, which helped relieve some of the accumulated stress from all the hard work we carried out during the day.

Some of the pieces I had to design were quick-disconnect protectors, part of environmental control that would pump oxygen into the life support system. Oxygen in its purest form is extremely flammable, so the absolute minimal amount of spark, say from hardware failure or leftover aluminum chips from the manufacturing process, will cause an explosion. I had to make sure these assemblies were protected from foreign object debris (FOD) to prevent the spacecraft from blowing up.

In the previous job I'd had at GE Aviation, the pace was much slower, and I didn't have to move around as much as I did here. Now I was juggling a wide variety of tasks: from designing, testing, building, and qualifying parts to navigating different sections of the spacecraft. It was all incredibly expedited, and I have my direct manager to thank for helping me understand how to handle and own my engineering assignments.

Leslie taught me how to go through the different processes, manage the varied deadlines, and navigate the many labs and relationships to get parts done by specific dates. It was the first time I owned two hardware assemblies that would eventually be manufactured and placed on one of these spaceships, and she guided me and helped me understand all the moving pieces from early on. In addition to Leslie, I sought out other mentors in different disciplines, including production engineers, rocket engineers, and technicians, who opened my eyes to the end goal behind my hard work. Then there was also Tien, a Vietnamese build engineer who taught me about his process and more about the rocket engines, how to refurbish components to make them reusable—which is what SpaceX is known for. Tien ultimately got promoted to lead the whole rocket build.

Aside from my learning invaluable skills, this work experience opened my eyes to a reality of working in STEM fields that I hadn't quite experienced yet: the lack of diversity. Although I knew that this was an overriding issue in our field, coming from a diverse high school and living in my diverse U-M and SHPE bubbles had sheltered me from facing it firsthand before my time at SpaceX. I spent four months as the only Hispanic intern among all the engineering teams and had met only one full-time Hispanic engineer. Women didn't abound either. With a few exceptions, such as Leslie and Tien and a handful of Indian and Asian guys, working at SpaceX was like entering a mostly white tech-bros frat house—I don't remember crossing paths with a single Black engineer. I did my best to fit in, play the tech-bro part, and accepted invitations to join my colleagues to hang out, but despite them being cool people, I often felt out of place. In the sea of white dudes with their beards and wide array of beer knowledge, I was beginning to be haunted by imposter syndrome. I'd often ask myself, *Wait, what am I doing here?* Since it was hard for me to build strong relationships with my immediate white peers, I turned to the lab techs, the welders, the manufacturing techs, who were a more diverse group of people I could easily relate to, and they seemed to demonstrate more empathy. But what I suffered through the most was the feedback I received from the higher-ups. I was suddenly made aware that some of my strengths, such as passion in discussion, were perceived as me being "too outspoken." At times I wondered if I should even be there. Maybe I just wasn't a good fit. It was the first time I really felt what it meant to be a Hispanic engineer in a STEM field.

Seeking some familiarity, I reached out to the University of Southern California SHPE chapter, and they took me in with open arms. I hung out and partied with them on weekends and even got involved in some of their SHPE activities, which didn't just help ground me; they also allowed me to fall in love with California.

I found the lack of representation at SpaceX so disturbing that when my internship was over, I wrote Elon Musk a letter expressing my frustration. Why did his teams so evidently lack diversity, equity, and inclusion? I really wanted to know. I would've liked to help fix that problem. Did he get my note? I don't know. Did he read it?

I'm not sure. Did the environment change over the years? I heard it improved somewhat, but judging by recent comments posted on Glassdoor.com, the company still has a lot of growing up to do in this particular area.

By the end of this work experience, I had an invaluable skill set and a clear idea of what I needed to home in on at school once I returned. All I wanted was to get back at it, wrap up my undergrad degree, and start grad school, but I had one more internship to go: Boeing, in St. Louis, Missouri.

Leaving Los Angeles, the beach, and that gorgeous weather for St. Louis wasn't easy, but my curiosity about what it would be like to work at Boeing was stronger than the pull I felt to stay in sunny California. When I joined the group, I was placed with their Direct Attack Weapons team, which develops smart missiles and JDAMs (Joint Direct Attack Munitions), another type of guided missile. This team also specialized in developing a subsystem and wing system that could be attached to leftover "dumb bombs" (which were unguided bombs dropped from planes and just exploded wherever they happened to land). No longer dumb bombs, these modified units could then be dropped over specific targets, and small wings would deploy to guide them to those targets rather than allow them to land anywhere in the vicinity. This meant that the United States went from having hundreds of thousands of leftover dumb bombs from the Cold War to usable smart bombs for targeted missions, thanks to this elegant mechanism. That was what I was tasked to work on during my fourteen-week stint at Boeing.

The first day was like at any other internship: introductions, orientation, training, getting my badge, and meeting my manager. I knew the drill by now. The manager invited me to Boeing's St. Charles site, just to check it out, and introduced me to the team there. I'll never forget one of the main electrical engineers: Darko Ivanovich. What a great name! I started calling him "the Villain." As I walked farther into the facility, I tried to curtail my shock. I had gone from a rocket to a missile factory practically overnight. The engineering building was near the storage facility that housed all the missiles and explosives. When I walked around the building, I quickly noticed a long yellow strip on the floor that divided the safe zone from the unsafe zone. In other words, if something exploded in the facility next door and you were on the wrong side of that line, you could potentially be toast.

An Air Force base is on-site, with Missouri National Guard present—this is where big military aircraft transport all over the world the products produced at this facility. Additionally, Boeing has their own line of jet fighters at this site, so they pump out airplanes from this factory and take them on flight tests. That meant every day, driving to or from work, I would catch a glimpse of F-15s or F-18s swooping by. It was like watching the HISTORY channel through the car window.

I had fourteen weeks to work on the different projects assigned to me, which wasn't much time given the scope of the jobs. Fortunately, I was using the same design software I had used at SpaceX, so I was equipped to jump right in and start designing, generating views, and working with drawings. The full-time staff was welcoming and helpful, but with only one woman in an entire team made up of mostly old white men, their diversity situation was even worse than that of SpaceX. Thankfully, we had a touch more representation among the summer interns in my immediate group: two Black men and another Hispanic man—sounds like the setup for a "walked into a bar" joke, right? The rest of the two hundred or so interns were mostly white and also well equipped with beer knowledge.

While I never suffered discrimination during my time there, I did see firsthand why diversity and representation matter. I was just five days shy of my last day at Boeing when Michael Brown was fatally shot by a white policeman in Ferguson, Missouri. I was there when people said, "Enough is enough," and took to the streets to protest. And yes, there were riots too. Along with numerous other businesses, places I had eaten at, like the Ferguson Brewing Company, were burned down during those weeks of unrest. But I was very aware of the social dilemmas that had led to such a reaction. It wasn't unwarranted. The nation seemed loud with calls for justice, and yet back at work, most people met the moment with silence, never mentioning the shooting or the injustice. If they said anything, it was about how the protests were uncalled for and the highway blockages were a nuisance. Suddenly, as I began to recognize my own presence in these spaces, I was immersed in the white perspective, and it wasn't pretty.

As much as I enjoyed working at Boeing, and as much as I appreciated being exposed to some amazing engines and cool technology, I couldn't see myself in that environment for the long haul. My internships that year didn't just equip me with invaluable work skills; they also painted a clearer picture of what I wanted and what I wouldn't tolerate out in the real world. It's not just about the job for me; it's about the people, the culture, the environment—a lesson that continues to be a driving force in my job choices to this day.

When I began college, I devised a system of "time blocking" for myself. Years before, when I was seven years old and had inherited a PlayStation from Xavier, it became apparent that I could spend endless hours on devices, potentially neglecting my schoolwork. So Mami, ever resourceful, grabbed a piece of paper, jotted down the days of the week, and, beneath each one, dedicated a block of time to each of my activities: school, homework, meals, exercise, sleep. Reviving her strategy over ten years later removed an extra layer of stress during my college years by providing me with guidelines I could turn to when my overflowing mind wasn't sure what I was supposed to be doing on a particular day or at a particular time. As I set my eyes on graduation and entered the next phase of my academic career, time blocking was a key resource, but it wasn't enough to keep my life balanced.

When I was in school full-time, my mind switched to super-brain mode. Aware that I needed to get a 3.0 to even be considered for a job—GPAs are a filter used by many established companies due to the sheer volume of applicants they must wade through—and a 3.5 for competitive grad school applications, I became hyper-focused on my coursework. In my final undergraduate year at U-M, driven by the real-world experiences I'd had in my internships and what I needed to know moving forward, I decided to enroll in four technically intense and brutal classes. Academics became my life. I wasn't doing any exercise other than a brief time-blocked walk, and any ounce of free time I did have went to my extracurricular work with SHPE. I was also a tutor and ran a supplemental mechanical engineering class on Sundays, for the same dynamics class I had loved so much prior to leaving to Los Angeles and St. Louis. However, balance is as essential as efficiency. As responsible and overworked as I may have been, I time-blocked Friday nights for pure fun. Those were the few hours I allowed myself to let loose. I'd meet with my buddies at Charlie's, our favorite local bar, dance it up at Rick's or hit up a party, and end the night at Fleetwood Diner, crystallizing some of my most important friendships to date in the process. To this day, I go back to those nights as a reminder that I can't engineer my life the way I engineer machines. I can't forget to punctuate my career path with some regular old fun for my personal well-being.

I already had two job offers on the table, but I continued to pace the SHPE National Convention floor in Baltimore, and that's when I saw NASA's Jet Propulsion Lab booth. Familiar with the rovers they had sent to Mars but unaware of where the projects for the next rover stood, I approached them and chatted with Eric Aguilar.

"I'm interested in controls and do a bit of coding," I said, explaining what I had worked on in the past. "So, what are you guys up to with Mars 2020?" I asked. Mars Science Laboratory (MSL) had landed a few years prior, and I knew that Mars 2020 was their next big project. There were some other smaller projects kicking off at the time too. It all sounded interesting, so I asked him what they were hiring for. Eric was the manager of the testbed team for Mars 2020.

"There's really great work coming along. We're early in the testing campaign for Mars 2020. And there's a great opportunity to work full-time."

"Okay, sounds good," I said, hopeful.

"I'm going to have you talk to Magdy," he replied and called this other guy over.

Magdy Bareh was one of the Mars 2020 big shots. I got a similar spiel from him, and as we continued talking, it became clear to all three of us that I was the right fit for what they were looking for.

"I'm about to graduate with my undergrad degree, but I continue at Michigan next fall on my master's," I said, bracing for impact.

"You don't have to worry about the cost, because I'm not only accepted but fully funded. All I need from you is to let me know that you'll accept me as a temporary full-time employee now, then let me go back to school for a year, and afterward I'll return to JPL full-time."

"Wait," said Eric. "You're telling us that you'd come work for us, get your master's, JPL doesn't have to worry about covering you, and then you want to come back to JPL?"

"Yes."

"That's not a problem at all. Why would that even be an issue?"

Within a few weeks, I was set up as a JPL grad-student intern with a January 2016 start date along with a full-time position waiting for me on the other side of my master's degree. As I digested this new turn of events, it hit me that I was being invited to work on building up the testbed that would be used for the Mars 2020 mission. As a kid, I had been mesmerized by the efforts to explore other planets and one day potentially find life, and now . . . *Holy crap, I'm actually going to be working on a rover!*

CHAPTER 12

CRUISE PHASE

**Mastering Engineering While Monitoring
a Family Reality**

By January 2016, I was officially ready to begin my mission to Mars. I would spend the next eight months in Los Angeles at JPL before starting grad school at U-M in the fall. Upon arrival, I had to sort out housing and transportation. Since I would be staying in Los Angeles for less than a year, I didn't feel it was worth it to shell out any money for a car. I relied on Zipcars for errands and exploring the city. And within days, I bought a bike, which I used for the seven-minute rides to and from work, since I had lucked out and found a room in a house located right outside of JPL's east gate.

Those bike rides reminded me of sophomore year of high school, when Serge and I had mapped out how long it would take to bike to school from our respective homes in Brooklyn. When we realized it took the same time as public transportation, we decided to try it out. We left our houses at 5:30 a.m., met up somewhere in Flatbush, then breezed down the traffic-less streets and over the

Brooklyn Bridge as the sun rose behind us. By 7:00 a.m. we were in the Lower East Side, saying hi to our classmates as we entered the school building. We thought we would make it a habit, but as the summer transitioned to fall and the temperature began to drop, so did our commitment to the bike commute.

I missed my friends from back home, but we had managed to continue seeing one another during our summer and holiday breaks and visited whenever we could, so our connections remained strong. Thankfully, LA was familiar territory, as I had spent several months there during my SpaceX internship, and I had some of my old friends in the vicinity, so escaping the crude Michigan winter for sunny California sure didn't put a dent in my days.

Those first few weeks at JPL were all about getting set up and trained to become a productive team member. I had everything from clean-room training to instructions on how to handle critical items, like flight hardware—hardware that would be placed on the Mars spacecraft destined to fly through space. Safety protocols for how to handle myself in these labs were also important, such as how to ground myself while dealing with electrical equipment so that if there was some kind of short, the current would be redirected through a wrap I would wear around my arm, avoiding potential harm or even death. Also on the list was heatstroke prevention while working in the Mars Yard—an outside area carved out of the surrounding hills, simulating a Martian landscape, where we would test different prototypes.

When I first started, there was only one other full-time person and a few contractors in our group besides our two managers. Our main task over the next few months was to convert the leftover testing ground support hardware that had been previously used for Mars Science Laboratory (MSL)—whose rover, *Curiosity*, had landed on Mars in August 2012—so that we could reuse it for our current mission. The Mars 2020 mission's architecture and much of the hardware that would be on its rover was designed to be similar if not the same as what had been used on MSL. By reusing as much as possible, we could dedicate our resources to adding new instruments and buying newer hardware where needed. This extended not only to the actual vehicle but also to the testing methods and campaigns, which are timelines of events and objectives used to establish confidence in the system's design. The larger scale issue no one on the team foresaw was that in assuming we would be reusing the previous mission's overall architecture and method, less money was budgeted for this work, which translated to less staffing. It would've been okay if the process had been as streamlined as JPL believed, but we soon realized that some of the old procedures didn't make sense for this new mission, so the workload and the pressure we felt swelled.

When we were developing and designing the Mars 2020 hardware, it was grouped into several tiers, and as we moved from one to the next, we inched closer to the final flight hardware that would be launched into space. We started off with prototypes—basic or printed circuit boards with test wires and components. Then we moved on to breadboards—hardware with increased fidelity and closer to the functionality of the flight hardware. Once we began to solidify our design with the breadboards, we were able to create the engineering models, which function like the flight design. However, the engineering models don't have to be built to support the violent shaking a rocket launch experiences or the radiation the flight-grade hardware may experience in space. Their functionality is intended to be the same as the final flight hardware but without the additional components needed to survive the extreme space environment.

This is all a result of years of trial and error, deep problem-solving, and extensive collaboration among colleagues.

The final design was the flight hardware, built to support all the dynamic loads on a rocket as well as any violent event, such as entry, descent, and landing, to absorb or withstand shock. It also had to survive the extreme temperature fluctuations experienced in flight and during the mission, as well as be radiation hardened—that is, able to withstand high levels of radiation.

Flight hardware is much more expensive than the previously mentioned hardware tiers; that's why we build up to engineering models, which often serve as our testbeds. In other words, if our flight hardware has issues in space, we can potentially replicate it with our engineering models on Earth and keep the flight hardware as pristine as possible. The models are used to do strenuous testing that we wouldn't put the expensive flight hardware through. If we find issues in the engineering models, we must consider whether we may see them in the flight hardware as well. It's a pain, but since we can't send a mechanic to space yet, with engineering models we are able to take things apart, fix components or replace them where needed, and physically update the flight hardware while it's still on Earth.

Once it's all launched into space, if we come across unforeseen issues, it's not the end of the world (or so we hope), but we can no longer physically modify the hardware. The only way we can develop workarounds or patches at that point is through software or through the way we operate and restrict usage of the hardware in space. For example, we may realize that we can't use a certain instrument while driving the rover, so we must add a mission constraint to indicate that this instrument should remain unused while the vehicle is in motion.

Subsystems are grouped circuit boards placed together in boxes based on what they correspond to, like the rover's motor controller, its computing element, its communication system, or any of its instruments, to name some examples. When we started plugging these subsystems together, we were able to see different interactions we weren't necessarily able to observe before, and that's when we started to find potential issues in hardware–software interactions. For example, what we were able to use in parallel or not, and what could generate electrical noise and affect another component. It was like running an operating system to start up your favorite video chatting software while also enabling the use of your camera—all separate processes that use separate subsystems within your computer. In other words, when we started integrating these subsystems, we were able to find the bugs living across the different data paths and copper paths. The objective was to verify that all designed behaviors worked as intended—which we call "verification and validation," a phrase we use *a lot*—and to find and fix as many bugs as possible before launching the vehicle into space.

If we used the wrong equipment to take specific measurements, we could easily damage parts, so when in doubt, I always double-checked with my co-workers. There are no dumb questions, especially when dealing with such expensive hardware where plugging in one thing incorrectly could break it. Thankfully, we always work in twos to verify each other's work.

Aside from the testbed work, I was asked to participate in a fun side project: a smaller rover, aka *ROV-E*, intended for outreach purposes, envisioned by JPL's director at the time, award-winning electrical engineer Charles Elachi. During his sixteen-year tenure at JPL, Charles Elachi oversaw the launch of twenty-four missions, and prior to that, he had pioneered the development of space radar to take images of Earth and other planets. Charles also had chaired committees that developed maps for NASA's exploration of our solar system, other solar systems, and Mars. Now I was tasked with getting the *ROV-E* in working order, not just for outreach but also for this engineering giant's retirement party. That meant I had six months to iron out all the kinks and get the small rover operational. I welcomed this challenge because it allowed me to step away from the electrical engineering I had been applying to the testbeds and scratch my mechanical-engineering itch.

First, I took some of the knowledge from my Design and Manufacturability course from a few months back and redesigned the motor hubs, the battery holder, and the inside of the rover's head so that it could correctly hold the cameras that would transmit a live video feed as it was roving. I also built tool kits with instruction cards and put them in small bags for the outreach team, so they had quick access to how to fix anything that might break during a demo.

By the time Charles Elachi's retirement party came around, I was ready to roll with *ROV-E*. My manager had built a tiny house and ramp for the occasion—that's where we hid *ROV-E*. Once the ceremony was underway, on cue, I hid backstage and remotely drove *ROV-E* out of its little house and down the ramp, interrupting Charles's speech midsentence. Then the voice of the little robot, from Adam Steltzner—Mars 2020's chief engineer for the entry, descent, and landing vehicle, and one of the main people to come up with the idea of using a sky-crane maneuver to land on Mars— proceeded to have some fun banter with Charles. Finally, I drove it back up the ramp and managed to make it raise one of its legs and wave. The presentation was less than a minute long, but the crowd (and Charles) lapped it up. Later that evening at the gala, there I was, the only intern, surrounded by aerospace big shots, including renowned engineer Bill Nye and SpaceX's own president Gwynne Shotwell. No one could wipe the smile spread wide across my face. The night was a definitive one in my life, as the gravity of where I was and who I was becoming began to sink in. I was standing among the giants of my field and was on the road to hopefully someday become one of them. What had once been only a dream was becoming my reality. I listened to people praise Charles for his long list of accomplishments while eating cookies with his face on them and hoped that one day I'd have a similar gathering celebrating my successful career, maybe even with my face on a few cookies.

As the months passed, I continued to have more moments when goals were realized and dreams came to fruition. Although I no longer held an official title at SHPE, I remained involved in events, such as the regional conference, which was held at Arizona State University that spring. The conference included leadership workshops tailored to undergraduate students and served as a great networking opportunity for chapters across their respective regions. This time around, I decided to focus on creating a conference-track curriculum that would engage graduates like me, who had already attended countless conferences, done the networking, and knew what to expect. I invited technical speakers to talk about current topics in their fields. We had a drone expert, a coding boot-camp director, and a top technology firm employee touching upon leadership, big data, and artificial intelligence. JPL also had a booth at the conference's mini career fair, but this time I stood on the other side of the table, helping recruit interns for roles that could change their lives. For the first time my outreach and professional worlds collided, and I had the opportunity and the honor of continuing to pave the way that others had paved before me, lifting as I climbed.

That summer, Mami, who had moved back to Florida with Xavier after Niní passed away, began to develop stronger anxiety and panic attacks. Up until then, they'd been much more sporadic and we had attributed them to certain stressful periods in her life, such as the years of struggle to make ends meet or even the recent loss of Niní. But when I flew down for a short visit during my birthday in April, I realized just how much more erratic her behavior had become. She walked around the house, closing all the blinds to prevent anyone from looking in, she refused to go anywhere, and her nights had become increasingly restless. Xavier wasn't sure what to do. They had not yet set up her Medicare and were somewhat lost and overwhelmed when it came to applying for federal and local aid, so I began to consider the option of moving her to Michigan for the year to live with me.

I crunched the numbers to see how much I would be making a month with the graduate student instructor position I'd taken and what my budget would need to be to support my mom if it came to that. Could I get an apartment for us, food, cover any weekly expenses, and still be okay? Yes, I could. Would I still be able to save enough money to pay off my student loans after graduating? No, but I knew I could handle that later. Could I live with my mom again after five years? Of course! She had been my shelter for so many years, now it was my turn to be hers.

I suggested the move to her as a way that I could help sort out her insurance and other health-related concerns, and she gratefully accepted. I explained my situation to my initial roommates, and then found Mami and myself a one-bedroom apartment that was only a five-minute bike ride from North Campus, where all my classes would be held that year. I wasn't at all fazed about having to share a room with my mom; we'd done it before. I did warn her that I wouldn't be home much during the week—as soon as classes started, my super-brain mode would reactivate, and I would start living and breathing my coursework.

On our first day in the apartment, thanks to one of my friends, we already had a big mattress to sleep on. We then set out to the Salvation Army to buy a few basics, like a makeshift dining table and a couple of chairs. I also used that time to begin familiarizing her with the town and its layout. As we made the place feel a little homier, my mom pulled out two framed pieces of art that used to adorn our walls in the studio apartment back in Caguas, Puerto Rico. When she hung them on the wall, it felt like I was traveling back in time.

By the start of my semester, we were settled in our new home and her Medicare and other aid, which I had sorted out thanks to a local health center, began to kick in. From then on, she was able to seek help for her anxiety and panic attacks without stressing about a bill that could leave her with a crippling lifelong debt.

Now that everything was on track with my mom, I started zeroing in on the semester ahead of me. That year I decided to double-dip between two master's programs: the one I had chosen, Integrative Systems and Design (ISD), and Climate and Space Sciences and Engineering (CLASP). Once I fulfilled the required classes, ISD allowed me to customize my master's, which meant I could take courses from other related programs. After spending eight months at JPL, I knew which ones might benefit me more and what I needed to brush up on, so I took advantage of the system and made it work for me.

As if my course load wasn't enough, I was also teaching Design and Manufacturing for juniors as a graduate student instructor (GSI). The professors drove the main lectures, while the GSIs designed the labs, including the contents, lectures, and exercises that went with them. I had around thirty students under my tutelage. Office hours were also part of the deal, to help students with their homework and any projects they had to build by the end of the semester.

There's a certain tone that appears in my voice when I'm teaching that commands attention and respect. It comes from a place of technical authority to speak on topics with confidence, but some students interpreted it, and me, as somewhat intimidating. Yes, I was a tough teacher. I called them out when they didn't do an assignment and was also extremely specific about what I wanted to see in their technical communication, because after my different work experiences, I finally understood why my professors had put such an emphasis on this portion of our classes. Being able to communicate clearly is half the battle in a real-life mission. On the other hand, I was also lenient and fair when it came to grading and constantly emphasized that I welcomed honest feedback on my own work with them. The memory of being belittled in class by a few professors for not being up to par with what they expected was still fresh in my mind and something I did not want to repeat. I aimed to create an environment where students felt free to speak up, ask questions, and challenge me. I wanted to improve as a teacher and make sure I did my best to help them understand and digest the topics we covered.

While I was always eager to listen to them, I sometimes stumbled with my empathy skills. My open-door policy once drew in a student who felt comfortable enough to tell me that she believed I sometimes was too tough on the class, and she wished I would take more time to stop and listen before providing feedback. I was so glad she brought this up—listening is a skill that I am constantly working on to this day. I've learned that while I'm often champing at the bit, eager to contribute what I know, I can take a beat and do my best to not get ahead of myself and others in a work or learning environment.

I made a point to come home for dinner whenever I could to share a meal with Mami, just like she had done when I was a kid and she was working several jobs to make ends meet. Mami seemed to be adapting well to Ann Arbor. Sometimes she'd wait for me with a box of pizza from around the corner. We also spent time together on weekends. That's when I would check to make sure she was going to her doctor's appointments and keeping up with her mental health.

Meanwhile, she started walking through town to become familiar with the area, and she eventually found a Catholic church in downtown Ann Arbor where she began to build her own little community. If weather permitted, she'd do the hour-long walk there and back, a pleasant stroll until the frigid temperatures arrived. Then she'd either take the bus or request an Uber. I admired her for putting herself out there in an attempt to defy the unpredictable anxiety and panic attacks that at times managed to paralyze her.

As the months progressed, so did her anxiety. She'd often wake up in the middle of the night, heart racing amid a full-blown panic attack. One time she climbed out of bed to shake it off and tripped on the way to the living room. I woke up to a loud thump, rushed to her side, and was relieved to see she hadn't broken anything. After a while, I got used to her getting up in the middle of the night, quietly moving over to the couch, then eventually returning to bed. But when I had an out-of-town conference to attend and couldn't bring her along, an undercurrent of worry besieged my days. To keep my calm, I'd call her a few times a day to check in and make sure all was running smoothly back home. Thankfully, I also always had a friend who would willingly step up and spend those couple of nights with her to make sure she was okay—the kindness of chosen family members was always strong.

Somehow I was able to strike a balance between my workload and the cloud of worry that hung over my head regarding my mom, but my self-care fell by the wayside. I stopped working out but was still as voraciously hungry as when I was regularly exercising and lifting. There was too much to do, too much to take care of to put my needs first. When things get rough, I do everything I can to fix it, oftentimes neglecting my own mental and physical health. It's fair to say that I have a productive approach, but turning toward work while neglecting my well-being is a machine-like behavior focused solely on outputs, and that has led to burnout on more than one occasion. Taking care of my mom pushed me to excel further in school and as a GSI. This wasn't just about making her proud; it was about making sure that I had the skills to land a well-paying position after graduating.

Mami had busted herself to provide for and protect me, to give me a better life. She also had always taken into consideration the needs of her students, her co-workers, how she was perceived at work, constantly looking out for everyone but herself. This, together with expensive rents and low salaries that left no room for investments, had led to her never being able to build up her nest egg to ensure herself a good retirement. Xavier and I were beginning to wake up to the fact that, like many immigrant and first-gen kids, we were Mami's retirement investment. She couldn't be at the mercy of her minimal pension from Ecuador, or the small, early retirement Social Security check she now received for her years of work in the United States. Combined, this wasn't enough to give her financial independence. She was our responsibility, so I needed to kick ass in class to guarantee I could take care of the both of us from then on.

Instead of the freshman fifteen, by the time I graduated I had gained the grad-school twenty. I never had washboard abs or muscles by any means, so I was used to hiding my tiny love handles under hoodies or sweaters, but now even they began to feel too snug around my waist. Insecurities about my looks began to creep in—I didn't want to have to get anything larger than my regular medium. Pushing physical activity to the bottom of my to-do list was a habit I knew I needed to break if I was going to pull my weight—for myself and Mami—in the long run. It's still a work in progress.

I spent the day leading up to my graduation decking out the top surface of my cap with a photo of a rover on Mars amid a starlit background in the center, a photo on top of that one used in the NASA JPL logo, and at the bottom the M-STEM, SHPE, and CubeSat logos, symbolizing my support network and where I was heading next. The next morning, we woke up at seven and I excitedly slipped on my gown, put on my stole with the Puerto Rican and Ecuadorian flags, and proudly donned my customized cap. Then I turned to Mami, who was also dressed for the big day, but she shook her head and averted her gaze. Disappointment gripped my heart. "No, I can't go," she whispered, defeated. She couldn't bring herself to deal with the overwhelming throng of people at the Big House—Michigan Stadium. That was one of the few times I truly lost my patience with her. It hurt. I was angry. I wanted Mami to be there, to watch her son walk during his master's graduation ceremony. After championing my education throughout my life, I couldn't help but think, *How could you not have the courage to fight this for me?* If I could go back in time, I would've smacked myself and said, "Muchacho de mierda, what are you talking about?" I was too immersed in the momentous occasion to glimpse the bigger picture. Mami was traversing the peak of her mental health crisis, and she knew that being exposed to such a large crowd could potentially trigger an episode. The last thing she wanted to do was ruin my day. In a way, she was fighting for me, but I just couldn't see it at the time.

The following morning, which happened to be my birthday, I woke up to find Mami smiling next to her freshly baked and decorated dozen cupcakes with U-M's logo and a beautiful blue graduation-birthday cake topped with our planetary system and three candles, with the inscription, "For today, goodbye. For tomorrow, good luck. Forever, Go Blue!"

CHAPTER 13

TRAJECTORY CORRECTION
Preparing to Reach for the Stars

NASA first began to send orbiters to Mars back in the 1970s, managing to capture the first images of the red planet, as well as land the *Viking* landers, but it took twenty years to land the first wheeled vehicle on a planet in our solar system. On July 4, 1997, *Sojourner* was the first rover to successfully touch down on Mars. At twenty-five pounds and only twelve inches tall, it was what NASA would later dub a "microrover." Designed to last one week, *Sojourner* made it to eighty-three days, during which it took atmospheric measurements, studied rocks and dust, and managed to snap more than five hundred pictures of the Ares Vallis, a large channel that appeared to have been the bed of ancient flood waters.

Then came *Spirit* and *Opportunity*, twin rovers that landed on Mars in January 2004 at different locations. In the seven intervening years from when *Sojourner* had rolled across the Mars surface, NASA had managed to create and build these "Adventure Twins," each weighing 400 pounds and approximately the size of a golf cart.

Their planned ninety-day mission far surpassed expectations, as mentioned earlier. *Spirit* made it to 2010, when its wheel got stuck in sand, stranding it from further exploration. But *Opportunity* continued its surveying until 2018, which meant that when I began my full-time position at JPL, it was still roving the Mars surface. *Opportunity*'s adventure came to an end only when a weeks-long planetary dust storm encircled every inch of Mars. That summer I had gone to the Griffith Observatory for their once-a-month star party, where they host amateur astronomers who bring their telescopes for a public viewing of the planets and stars up above. When I put my eye against the viewfinder to see Mars, instead of its renowned overlap of oranges and reds, I saw a solid orange ball. The dust blanketed the entire planet, including *Opportunity*'s solar panels, impeding its batteries from recharging and leading it to ultimately freeze. RIP *Oppy*.

NASA had launched another rover to Mars, *Curiosity*, which landed on the Gale Crater on August 5, 2012. This was the first time a supersonic parachute, a jet-controlled descent vehicle, and the sky-crane technique were used to accommodate the landing of this much larger and heavier rover—2,000 pounds and about the size of a MINI Cooper. It's equipped with seventeen cameras and tools, and still active today.

The sky-crane maneuver came from an idea that started as a drawn concept on some napkins and has proven to be a reliable way to deliver complex surface-exploration rovers to Mars. The concept was inspired by heavy-lift helicopters, but the method relies on thrusters instead of propellers. And contrary to a heavy-lift helicopter, upon delivery of a rover, the crane flies away and crashes itself on the surface far from the rover. As Adam Steltzner said about the sky-crane design, "It was the right kind of crazy."

Now it was our turn with *Perseverance*, the rover we were scheduled to launch on July 17, 2020, the Mars 2020 mission. Our rover is like an upgraded *Curiosity*, weighing 2,300 pounds and similar in size to *Curiosity*. Aside from capturing spellbinding photos and essential data, one of its key objectives is to collect Martian rock and sediment samples in search of signs of ancient microbial life. When I joined the mission in 2017, NASA was in the testing phase, with still a ways to go before *Perseverance* would be fully built and functional.

The rovers *Sojourner, Spirit, Opportunity, Curiosity, Perseverance*—and in April 2020 the helicopter *Ingenuity*—all received their names through naming contests that required K–12 students to submit an essay explaining why their proposed name would best represent the new Mars vehicle. From 28,000 entries from across the United States in 2019, the winning essay, selected by NASA, was submitted by seventh grader Alex Mather. He noted that many of the previous rovers' names were qualities that we possess, traits needed to explore space, but we forgot the most important one: perseverance. "The human race will always persevere into the future." Thank you, Alex, for that resilient sentiment.

We still had three years to go before our official 2020 launch date, and they would prove to be three of the most intense years of my life. Since I had already done an eight-month stint at JPL, I hit the ground running when I returned that summer. This time, I brought my mom with me to California to stay for about a month. My older niece, Amber, flew in to spend the last week of that month with us and then take Mami back home to Florida.

By the time they left LA, I was already a week into my job at JPL. Technically, I was still an intern because I was wrapping up my practicum and report for my master's, which was due in August. But for all intents and purposes, I was a full-time systems testbed engineer. As I began to dip my feet into the work that would completely take over my life for the next four years, that spring I had the chance to see my first ever rocket launch live.

Not a cloud dotted the sky that morning in Lompoc, California. My friends and I embarked on the two-hour journey bright and early, excited to see SpaceX's *Falcon 9* rocket take off from Vandenberg Air Force Base. The Iridium 2 mission was set to deliver ten Iridium satellites to low Earth orbit, part of the intended seventy-seven Iridium satellites to provide mobile satellite service to a variety of handheld devices. My buddy Justin Foley, also on the testbed team, had been here many times for other launches so he knew his way around. He is an amateur astrophotographer and also takes footage of rocket launches. We settled into our chosen spot, relatively far away but with a clear view of the rocket. And then we waited. Finally the rumble of those engines reverberated through our bodies as they thrust the rocket into the sky. The moment lasted only a few minutes, but the roar of that rocket made a long-lasting impression and swiftly put my work into perspective. I now had a live image of what our rover would have to endure to survive the launch and simply reach space, not to mention the months-long journey to Mars. We had our work cut out for us, that's for sure. After the rocket disappeared into the depths of the clear-blue sky, we drove over to the nearby Martian Ranch & Vineyard for a toast to the memories we were creating and the long road ahead.

On September 20, 2017, Category 5 Hurricane María hit my beloved Puerto Rico, hovering over the island for the next forty-eight hours, uprooting trees, causing power and phone outages, and inflicting catastrophic devastation throughout the land. It was a terrifying stretch of time when those of us with loved ones in the path of this destruction could only hope and pray they were okay. As we waited to get any type of news, my fix-it mentality kicked in—I needed to do something to channel my helplessness into action. I joined forces with a Puerto Rican who worked in another team at JPL to begin collecting donations, so we would be ready to ship them out as soon as it was possible. Relief washed over us both when the worry-laden silence was finally broken and we heard from our respective families and friends. More than anything, they had suffered material damage to their homes and surrounding streets, but everyone within our circles was okay otherwise. Rosa and Sonia described the experience as a powered-on jet engine sucking everything up into the air. As more news was released of the extent of the damage people had suffered, my friend and I continued to organize donation efforts in Los Angeles. It was all we could do at the time.

I had to carry my worry while I continued to work. I was assigned to avionics and thermal functions testing. In simple terms, the rover has two brains: its main day-to-day brain and what I call its lizard brain. The lizard brain is always running in the background, ready for fight or flight. It checks to make sure that the main computer, or main brain, is working well. If something goes south with the main brain, then the lizard brain can go through particular states to keep the system at a basic level of safety, putting the rover in a partially autonomous configuration that allows us time to figure out what to input to safely reconfigure its hardware.

The rover's thermal behaviors are what helps keep it alive overnight, when Mars temperatures can drop to –100°F or lower, depending on the season. There are particular instruments and mechanisms that can only operate within a specific range of temperatures. If they become too cold, we must be able to heat them up. If they're too warm, we have to stop using them or actively cool them down to the range we want them to operate in. As we gradually entered an all-hands-on-deck phase ahead of our July 2020 launch date, I knew that if I was going to be an effective and successful member of the team, I needed to make the conscious decision to put my work first, but not before making my all-important pit stop to spend Christmas with my family.

This time we met up in Florida. My grandparents, who didn't travel often, joined us from New York. And I got to reunite with Sonia and Robert, who were temporarily living in the area while they sorted through Hurricane María's damage back home. While my abuelo made sure the TV and music were set up and ready for our gathering, my abuela got busy in the kitchen, whipping up her famous casuela or caldo de bola together with extra sides to keep us all fed, full, and happy. My tías and tíos would give them a hand while making fun of each other and roasting my cousins. And a round of Telefunken (a game similar to rummy) was always in order, with bets of up to two dollars per person per round.

The highlight of this break wasn't just spending quality time with my relatives and chosen family; it was also getting the chance to take my ninety-one-year-old grandfather and my brother to the Kennedy Space Center—a first for the three of us. Walking into the center and suddenly being in the presence of all this antiquated hardware took my breath away. The exhibit featuring the *Saturn V* launch vehicle made me feel so small. I was mesmerized by how the 1950s team was able to design the stunning hardware displayed before me with the limited technology they had access to in comparison to what we have now. Sure, they had a relatively bigger budget and thousands of people working on one problem, which is not a luxury we enjoy, but they didn't have our software and automated procedures, and they were doing it all for the first time. As if taking all of this in wasn't enough, being there as a NASA engineer, walking the entire center by my grandfather's side, with me as our tour guide, explaining each piece before us, was an unparalleled full-circle moment for me. I stopped several times, glanced at my grandfather, and quietly asked, "Abuelo, are you okay? Would you like us to sit down for a little while to rest?" but he outright refused any break, likely pushed forward by a sense of pride for his walking abilities as well as the sense of wonder that had taken hold of us all as we witnessed this history-making equipment. It was an unequivocal reminder of the legacy I was now helping build with the Mars 2020 mission.

Inspired by the history I had witnessed at the Kennedy Center, and with a renewed sense of purpose, I was more eager than ever to dive even deeper into the mission at stake. February 2018 found me interacting with the *Ingenuity* helicopter for the first time, more specifically its base station, a component of the helicopter system that would live on the rover. This is the piece of hardware that would communicate with the helicopter on Mars. We were developing the capabilities, the hardware, all of it, to fulfill a technology demonstration to test the first powered flight on Mars, but NASA HQ still hadn't given the okay to add it to the Mars 2020 mission. So we were operating with the hope this green light would eventually be given, and we kept plowing ahead on the rover side, considering how we'd carry the helicopter, how we'd communicate with it, how we'd operate it from this base station. Initially, many of the people on the integration side of the rover were against the idea of integrating the helicopter as a separate system, because that meant it would also have its own separate battery. What if its battery caught fire while cruising through space or on the Mars surface? How would that damage the rover itself? "There's no way the helicopter will work" was one line of thought. The other: "There's no way you'll be able to get all of this work done in time." And the third: "This helicopter will be a distraction from the rest of the science the rover has to accomplish." Was it a risk to do this tremendous amount of work for a helicopter that might never launch? Yes, but it was one some of us were willing to take.

As the summer neared, I set my mind on Puerto Rico and the risks and sacrifices they had been forced to take when Hurricane María hit their shores. The island had far from recovered from the damage sustained a little less than a year earlier, and my colleague (turned girlfriend) and I were still eager to help in any way we could. I decided to use my social media to reach out to teachers in Puerto Rico to see how we could help that summer. I quickly received a reply from a U-M friend whose mom had a colleague, Marisa, in need of some help. With the community's blessing, she and her husband had decided to take over an abandoned school in Los Naranjos, a neighborhood in Vega Baja, located near Dorado, and turn it into a community center. The local residents had lost so much during the hurricane that she was hell-bent on making a difference. Now they were looking for volunteers to get the center off the ground. My girl-friend and I created a three-day STEM program for kids between the ages of eight and fifteen, called Ingenieros del Futuro (Engineers of the Future). The activities we planned introduced the kids to basic engineering concepts and revolved around three themes: robotics, electricity, and rockets. I set up a GoFundMe to help pay for some of the materials, while we paid for everything else out of pocket.

When we arrived, seeing the devastation firsthand threw me off my orbit and momentarily pushed me into an impotent void. As I painstakingly drove through intersections where the traffic lights had gone dark due to the lack of power, I slowly took in the trees scattered around the area like giant twigs, displaced rooftops, cut-down electricity cables, and attempted to store this harrowing data in a corner of my mind so I could find my way back to our main focus: the kids. I'd give myself time to process this emotional oscil-lation later, when I returned home.

We immediately got the kids working and building several projects—a basic robot, an electric car that used a solar panel to power it, a satellite model, and a wind turbine—to illustrate robot-ics, sustainable energy, and space exploration. We also scheduled outdoor time to give their brains a break and burn some energy playing soccer with us. For the last project of their three-day jour-ney, I taught them how to build a rocket with a two-liter plastic bottle and a few other readily available components. I had also pur-chased a bottle launch system that pumped up the rockets and had a trigger that allowed each kid to send their own rocket into the air. Once it reached a certain height, a parachute they had built into their system with their own hands deployed, safely landing their creation. Their excitement during each launch, descent, and landing, about further engaging with technology and pursuing opportunities in STEM, gave me hope for the people of Puerto Rico. The island currently has to import most of its food, despite once being fully reliant on its own agriculture sector. With agritech becoming more accessible, combined with the development of hydroponics, vertical farming, and more, I see this as a potentially booming sector for Puerto Rico in the future. But they will need dedicated STEM workers to make it happen. The same goes for the ever-controversial power grid. As energy storage and solar, hydro, and wind power become more accessible, microgrids will thrive, and so will the jobs related to those renewable systems.

Sinergia Los Naranjos is still active in the community. Marisa successfully launched a kitchen for folks to run catering businesses, and her husband, Ricardo, runs a reef restoration effort where many of the kids participate and get scuba training. Workshops occur in partnership with local student groups from nearby universities, mostly through grassroots funding and efforts. These kids have the power to build a better future, and I hope to continue to be able to come alongside them and encourage these developments through outreach, philanthropy, and policy influence.

By the spring of 2019, I was working with a few team members to test the capability of our rover to charge the helicopter battery through its base station while traversing space. Batteries, including those in computers and cell phones, left uncharged for a long period of time lose their properties and can't regain their full charging potential. Similarly, overcharging a battery and leaving it stored for a long period of time will degrade its lifetime. We had to figure out the sweet spot for the helicopter battery, then find how to measure that charge and, based on that, how to charge it from the rover battery. Once we figured this out through tests and failures and finally verified what worked, we had to come up with the sequence of steps that needed to be taken to charge the helicopter while flying through space. It was a complicated set of tests that took up a lot of our time but was essential to the helicopter's functionality and safety.

That summer I began to write and execute integration procedures for the helicopter deployment system, which is the assembly at the bottom of the rover that would hold the helicopter and deploy it. The system consisted of a tiny robotic arm with a motor that would keep the helicopter upright so that it could be successfully dropped onto the Martian surface. After testing this capability and gathering the necessary parameters, we determined that we could indeed deploy it on Mars. A short while after this, JPL finally approved the addition of the helicopter to the Mars 2020 mission. We got the green light. Like most times in my life, the risk proved to be worth taking.

With fewer than 365 days till launch, we began to focus on the incompressible list of tests necessary prior to takeoff. One of the overarching system flight software behaviors that needs to be functional for any vehicle's success on Mars is called SFP (system fault protection). Most hardware is redundant: if a computer goes offline, another with identical functions can kick in while engineers figure out root causes of the issues. Meanwhile, the helicopter *Ingenuity*

was finally brought to the assembly facility so it could be integrated into the rover. Seeing this nineteen-inch-tall vehicle in the flesh was absolutely thrilling. It was built to be strong and weighed only four pounds, light enough to be stowed away under the rover during its trip through space. But its four-foot-long rotors had to be extra thin yet have enough surface area to lift the unit in Mars's thin atmosphere, which is less than 1 percent as dense as the atmosphere on Earth. If these rotors were spun in Earth's atmosphere, they would explode, because they're so thin and fragile. So its flight capabilities had to be tested in a special pressure chamber where the atmosphere could be pumped down to replicate the one on Mars. Then, to emulate Mars's gravity, a pulley system was attached to softly assist the helicopter during liftoff, so it would act as if it were in lesser gravity. These tests, albeit quite basic, because they could only move the helicopter up and down, helped the team build the software that would eventually fly *Ingenuity* on Mars.

When office space became available in the Spacecraft Assembly Facility where the rover was being built, JPL decided to move the entire testbed team there too. The first probes launched to the moon, Mars, and Venus had been assembled there, as well as all of NASA's Mars rovers, *Galileo* and *Cassini* (the first orbiters sent to Jupiter and Saturn), and the twin *Voyager* spacecraft. I now had a desk in the same building as these giants. This meant that I could exit my workspace, cross the hallway, and stand in a viewing room overlooking the floor where the team was building the rover that would travel to Mars. I started calling it my break room. I'd sip a cup of coffee while watching, with childlike fascination, the rover getting built below, in disbelief that I wasn't just a spectator but an active participant on this mission. This was also the first time I had my own cubicle, almost like my own little office. My decorations of choice: LEGO sets, of course, including the Apollo 50th Anniversary lunar lander set that had been recently released. Much like the first time I ever had my own room, back in my junior year of college, the rigors of work were so extenuating I was barely able to enjoy it.

I continued to conduct tours of the testbed, which I had started back in 2016, and now I was able to show my guests—usually friends or friends of friends—the only room in the world where three different Mars missions were being tested at the same time: Mars 2020, *InSight* lander, and the MSL testbed. I'd turn on the Mars light—which we used to simulate the planet's lighting conditions to test our cameras—to show my guests how Mars would look to our own eyes if we were to stand on its surface one day. Because of the way the light hits the atmosphere, everywhere would basically be tinged with an orange hue. Then I'd take them to the Mars Yard up the hill, which at the time was being expanded with a bigger shed that would eventually house both MSL and Mars 2020 vehicle system testbeds, the Earth versions of their Mars counterparts. I had fun during these tours, in a way they served as a welcome break from the long hours of heady work.

As we entered 2020, survival mode started kicking in big time. For the past three years, I had been using kickboxing or jiujitsu as my stress reliever and main outlet. I would head over to the Fight Academy of Pasadena at six in the morning or late in the afternoon, depending on my work shift, two to four times a week. Shifts were based on the needs of the mission, which constantly fluctuated, and what needed to be accomplished to meet design gates and deadlines. This meant we'd have to push between five- and seven-day workweeks with two to three shifts per day. First shift ran from 7:00 a.m. to 2:30 p.m., second shift from 2:30 to 9:00 p.m., and third shift from 9:00 p.m. to 6:00 or 7:00 a.m. Depending on what was needed, many of us were often asked to stay on for a third shift. This meant a sleepless night followed by forcing myself to squeeze in some rest during the day. Furthermore, since the shifts weren't fixed, we never knew when we might have to stay overnight, which completely threw my sleep schedule into a tailspin. As the launch date approached, it was impossible to predict when I might have some free time, and often I had to use whatever slot I could to catch up on sleep. So once again my physical well-being, and in this case my jiujitsu classes, began to fall by the wayside.

Operating normally on this grueling schedule was not an option. I had spent my college years time blocking to avoid pulling all-nighters, so working through the night was completely new to me. Suddenly my life no longer revolved around time management but around priority management: What did I have to get done immediately to make the launch date? That's where our thoughts were back then, and that's when I slowly began to neglect myself and my mental well-being. My relationship with my girlfriend had ended only a few weeks earlier, on New Year's Eve, and it had left me swimming across an ocean of emotional wreckage. It was a confusing, volatile time in my life. Bone-tired, I did my best to summon what little fuel I had left in me to gear up for the increasing pressure at work, while also trying to get my heart and mind in order.

Even still, those long and harrowing weeks were punctuated by moments of pure awe at what we were working so hard to accomplish, like the day I caught a glimpse of *Perseverance* being packed up and readied for transport to Cape Canaveral, taking it one step closer to its forthcoming launch. I couldn't believe we had already made it this far and yet we still had so far to go. I latched on to that feeling of bewilderment and turned it into a propellant that would push me down the rugged road ahead. And then, seemingly out of nowhere, the COVID-19 pandemic arrived and the city, and our world, was on lockdown.

What will happen to our mission? Will we be able to hold the launch date? How will we move forward when life has come to a screeching halt? It took the Mars 2020 team about a week to adapt and come up with some clear guidance, but with the rover already at Cape Canaveral, there was no turning back; we had to finish what we'd started. A group of us were deemed "mission critical," which meant that although we did begin to work remotely, we could also access the lab when needed. We had to quickly convert our hardware and procedures so we could test remotely, and we equipped ourselves with masks, face shields, and gloves for when we did have to enter JPL.

Two months before our moment of truth, SpaceX launched astronauts to the International Space Station, the first time this had happened on US soil since the Space Shuttle fleet was retired in 2011. These headlines mixed with the Black Lives Matter protests kicking off across the country. I couldn't help but recall the images of the *Saturn V* rocket while the Civil Rights Movement was in full throttle on the ground. The similarities were hair-raising. Clear technological progress had happened in the last fifty years, but not much had changed in the civil rights department. Again, I found myself thinking, *I can't believe we've made it this far and yet we still have so far to go.*

After the long, arduous hours of work for months on end, and experiencing a worldwide pandemic lockdown while we had to power through, the day was finally upon us. I flew to Orlando, where I met up with Mami, whom I hadn't seen since January, and then we headed over to my friend Ryan's place; he had generously opened his home to host everyone in my group. Close friends from high school and college as well as co-workers were flying in from across the country for this momentous occasion. The idea of seeing some of my nearest and dearest—including Adisa; one of my best friends from college, Uzoma; Xavier; Ruth; my mom—after an imposed hiatus and on such a crucial day filled me with immense and uncontainable excitement. Amid the chaos and turmoil of the ongoing and relentless pandemic, they came through for me that day, ready to celebrate this massive accomplishment by my side, further cementing that they were my forever crew, always willing to go the extra mile to be present.

The day before the Mars 2020 launch, we rode into Cape Canaveral, and as we were driving around the area, on our way to see our *Atlas V* rocket, I passed the SpaceX site, took a wrong turn, and stumbled upon another familiar rocket. A few weeks prior to this momentous day, a group of friends and I had decided to drive two hours north of LA to Mount Pinos, to a specific location—once again guided by my colleague and friend Justin Foley—to see the flyby of the SpaceX Demo-2 mission as the capsule returned to Earth with astronauts on board. And now here we were, in the presence of the *Falcon 9* rocket that had put those very astronauts in space—a significant moment because it had been the first time astronauts were launched from US soil since the last component of the Space Shuttle program took flight in 2011. Heart racing, we hopped out of the car to take a closer look. Its sides were charred, but it was still in one piece. Seeing this part of aerospace history took my breath away. I needed a minute to take it all in. Even though my contribution at

SpaceX a few years earlier was likely equivalent to a grain of sand, I had still been a part of this endeavor, and the mission had been successful. The enormity of it all was overwhelming.

After climbing back into the car, still shaken by this happenstance, we kept driving until we reached the *Atlas V* rocket, with *Perseverance* attached to the top. A whole group of us from work gathered at the launch site and snapped an amazing set of photos. It was the first time we had seen each other in person in months. I stopped and took a good long look, knowing this would be the last time I would be that close to the rover we had spent years putting together before it hitched its one-way ticket to Mars.

The next day, eager to find a good spot on the beach to see the 7:00 a.m. launch, we decided to arrive two hours early. As we killed time, I started to joke around and said, "Listen, if this rocket blows up on the launchpad, that's it, I'm quitting and going into the farming business." We all laughed, but an uneasiness stirred within. Could the rocket that carried the baby we had been gestating for the past several years manage to make it to space, let alone to Mars? I quickly shook the thought out of my mind and focused instead on the morning sky, the beach, and being surrounded by Xavier, Ruth, my two nieces, their families, my friends, and of course Mami— the woman who had sacrificed everything to send me off was now steadfast by my side, watching my baby take off. A full-circle moment if there ever was one.

Meanwhile, as the time drew nearer, I was interviewed for several Spanish-language TV shows and networks, which I took on as an immense responsibility since there weren't many Hispanics who could speak of this mission. And then there was the crew from PBS's *NOVA*, who were there shooting the scene for their "Looking for Life on Mars" episode. That show had made such an impact on my life as a kid that when it was my turn to be onscreen as an adult, I was colossally aware that my words and message could potentially inspire someone else to pursue a career in STEM—maybe another Ecuadorian kid who dreamed of defying gravity—and I wanted to get it right. It was my opportunity to pay it forward. Nevertheless, it was hard to put into words how excited, nervous, and scared I felt in those moments prior to the launch. My buddy Jesse, who was also there, had a direct connection to the operations room, so after my interviews wrapped, we tuned in to the mission's status. Finally we heard the words we had been waiting for. "LD is a go . . ." This was it, the time had come! "You have permission to launch."

CHAPTER 14

ENTRY, DESCENT, AND LANDING
The Rover, the Helicopter, and Me

With only moments to spare, we ran into a problem. It was a frantic but fixable problem: we were all facing the wrong direction. Thankfully I alerted everyone just in time for us to scramble around and see our rocket soar up into the clear blue morning sky, leaving a trail of white smoke in its wake. My crew exploded into cheers and claps, pointing to the sky, while my mom nudged me to make sure I hadn't closed my eyes as we watched the *Atlas V* rocket take *Perseverance* on its first and last flight into space. I forgot to breathe for a few seconds as I saw it disappear into the universe, knowing that an hour later the rocket would release our spacecraft into a hyperbolic orbit, which would conclude its powered flight and send it on its seven-month cruise to Mars. Then I placed my hands on my heart and tumbled to the sand below in disbelief. This was a moment I had dreamed of my entire life. *We did it!* I thought, and then I lifted myself from the ground and joined my family and friends in what suddenly felt like a make-shift beach party teeming with ecstatic celebration.

A few years earlier, *Curiosity* had found evidence of organic molecules on Mars. Now it was *Perseverance*'s mission to gather samples of rocks in Jezero Crater for us to finally determine if the origin of the organic molecules on Mars is biological or rather geological, and answer the question that has haunted us for decades: Was there ever any form of life on the red planet? But first we had to get there.

Just like life had already taught me, the launch was only the beginning of that journey. *Perseverance* was slated to land on Mars on February 18, 2021. That left us with little more than six months to continue the next phase of testing and prepping for this milestone. While the cruise operations team focused on navigating through space, on the rover side we began to focus on operation readiness for surface activities. We'd grown used to operating the rover in our testbed, instantly receiving data, but now the paradigm had shifted. We would only be able to downlink data from Mars a few times a day, which we must then assess to understand the latest health status of the vehicle. And we had to plan activities around the single uplink window per day, which would be when we could generate and send commands to Mars.

In addition to helping anticipate how the rover's functions would work on Mars, we had to start testing the teams to understand how humans would operate the rover on Mars from Earth. What tools needed to be ready? What tools did the teams need to be trained on? These tools would allow us to interpret different data coming from Mars, so it was essential everyone assigned to downlinking information could also understand it. Meanwhile, I designed the setup procedure for the early surface operations readiness test. The objective of this role-playing test was to be as prepared as possible to respond to any anomaly we might encounter while the rover roved on Mars, including scenarios like what to do if our thermal operations engineer for the day got sick with COVID. The early surface operations readiness test would simulate the first ten sols on Mars.

Mars is the fourth planet from the sun. A day on Mars is known as a "sol" and it has twenty-four hours and thirty-nine minutes, as opposed to our twenty-four-hour days on Earth. It takes the equivalent of 1.88 Earth years, or 687 Earth days, or 668 sols, for Mars to take a lap around the sun. Gravity on Mars is about one-third of Earth's, so if you weigh 100 pounds on Earth, you would weigh about 38 pounds on Mars. This is due to the differences in mass and size between Mars and Earth. Mars is about half Earth's size measured through the equatorial diameter, and about one-tenth Earth's mass, so about 642 sextillion kilograms. To put things into perspective, you can fill Earth's volume with a little bit more than six Mars planets. Temperature on Mars can range from –284°F to 86°F, depending on location and season. Mars has two moons, Phobos and Deimos. All of these differences have to be considered when designing spacecraft and their respective concept of operations for entry, descent, and landing.

In order to emulate sol zero through five on the testbed, we worked overnight, so data could be downlinked by the respective operations team during the following Earth day and they could interpret it and plan the next day based on the status from the previous sol. Since we knew people would have to work whenever the rover was awake and whenever we could downlink the data, we planned the test with this in mind.

Fortunately, the first surface role-playing test went great. The teams began to get a grasp of where the holes were as well as what was working well in their processes and their respective ground tools, so they could move forward with confidence. Once completed, a few of us took a beach day to decompress. We brought a cooler and spent the entire day sprawled out on the sand, letting the ocean breeze melt our stress away, before we had to head back and keep going.

As if I wasn't doing enough, in those months leading to the end of 2020, I also began to work with the mobility subsystem team, which tested the autonomous driving on the rover. From the very beginning, I had wanted to understand how the rover made decisions when driving autonomously, so when I saw the opportunity to get involved, I asked if I could participate, and I was tasked with a specific function called the "leash." Like a dog leash, we could program the rover to go only a certain distance while driving autonomously. Meaning that if we programmed it to drive only twenty feet, it would come to a stop when it reached that mark. My objective was to figure out how to use this leash based on the rover's different driving odometers. This gave me great insight into what needs to be considered when driving autonomously on Mars, including how to drive around obstacles. I enjoyed the change of scenery—it was a little different from what I'd been doing the past few years, but just as relevant. It also gave me the chance to study new flight software code and new functions.

By December 2020, with only two months to go before entry, descent, and landing, I headed to New York for three weeks to work remotely and spend Christmas with my grandparents. Before leaving LA, I set up webcams in our lab, so I could have a visual of what was going on in there when needed. My colleague Anais was local to LA, so she would be at the ready if required, plus she was scheduled to run the next mission role-playing test, so it was a perfect time for me to take flight.

I spent those first three weeks of December hunkered down in my uncle's basement in Mill Basin with my computer and a projector that allowed me to have a second screen to run things remotely. The basement had been converted into a rental studio, a definite upgrade from when I had lived there and slept on an inflatable mattress in the living room. And this time around, there were no other guests living in the house; it was just Tía Jenny, Tío Oscar, and their daughter Melody. Between our crazy shift hours and the time zone difference, I was basically up all night, heading to bed as Tío Oscar and Tía Jenny left for work around five in the morning. Once or twice a day, I would mask up, leave the basement, and head to the twenty-four-hour Dunkin' Donuts down the street to fuel up with coffee and a snack before diving back into work. When my self-imposed social isolation ended, I emerged with a sigh of relief to spend time with my grandparents drinking coquito or mulled apple cider and catching up with the rest of the family and some friends. It was a heartwarming reprieve amid the strains of isolation, the endless work hours, and the accumulating pressure of *Perseverance*'s forthcoming landing.

As February approached, my next big task was to design the shadowing procedure for the early surface operations. Once we landed on Mars, the testbed would have to shadow the rover for ten days on the surface of Mars. For this to happen successfully, the testbed had to be a few hours behind the rover on Mars so we could be as close as possible to the state in which it could run into a problem. This way we would have enough time to configure and match whatever issue may have happened in flight and immediately try to debug anomalies.

Meanwhile, another testbed, designed specifically for the entry, descent, landing, and cruise operations, had been set up by Justin to run ahead of Mars time so we could also figure out if we might run into any issues and, in turn, fix them before they happened while the vehicle was cruising through space.

We coordinated handshakes between the testbeds and teams, so shadowing between both testbeds could be done as smoothly as possible. We were so focused on these tasks that we barely had time to ruminate on the what-if landing scenarios. Would Mars welcome us or destroy us? We were working our damnedest to anticipate all scenarios to make sure the outcome would be a positive one.

January 2021 was the month we revved up our outreach efforts to prepare for the big landing. The media coordination was handled by NASA's PR department, and I was really pleased because not only would we be getting coverage in English and Spanish but also, for the first time ever, there would be a whole mission broadcast in Spanish, Mars 2020 no less—what would turn out to be one of the events with the highest viewership that year! This effort was spearheaded by my mentor and manager, Diana Trujillo, and was yet another fantastic example of her leadership as well as how to push boundaries and remove obstacles for our Hispanic community.

It also meant my friends and family in Ecuador, Puerto Rico, and New York could sit back and enjoy the moment without needing anyone by their side to translate. On the flip side, this meant I had to squeeze in interviews during one of the most stressful, nonstop stages of the mission.

As if that weren't enough, only a couple of weeks before landing, the helicopter team began to put together the actual sequences that would allow them to release the helicopter from under the rover and deploy it on Mars, and they needed the support of a testbed engineer, so I was asked to lend a hand. Having worked on the testing of the helicopter's base station and deployment arm a couple of years earlier, I'd be able to share thoughts from my previous experience. They also needed to test how to interact with the helicopter and the Mars operations systems, not only with the rover but also on the ground. As I observed them at work, I couldn't believe we were already so close to when the real action would take place.

The shifts and schedules were tricky because we were starting to prepare ourselves to operate on Mars time, which meant every day would now end forty minutes after the twenty-four-hour mark. I also had to calculate how many engineers we would need for support, who would be in touch with the operations team, and how we would grab the information from the operations room and bring it into the testbed, all while making the transition between testbed engineers ending and starting their respective shifts as smoothly as possible. It was a lot of responsibility that I was honored to hold, but at the same time I felt the weight of it and couldn't help but think, *Holy crap. I hope I'm not missing anything!* Everything had to flow seamlessly, like a well-choreographed dance, for the health of the mission.

Then, in February, just days before landing, I was also asked if I could be a part of the operations team. It was clear we were understaffed, and I was beginning to stretch myself way too thin, pushing far beyond my limits; but there just weren't enough people to support the hours and the operations heading our way in light speed, and I was dead set on doing everything in my power to help us see our rover through its landing and crucial first sols on Mars.

I got to the lab bright and early on February 18, 2021. There was still work to do, but the excitement in the air was palpable. A group of us had decided we would be in the vehicle system testbed shed, at the Mars Yard, anticipating the landing. Anais and I were selected to be part of an internal panel for an employee JPL streaming of the big moment, with approximately three thousand employees watching from their homes due to the pandemic. As we prepared for landing, I began to get texts from friends: "Hey, I just saw you on TV!" NASA was live streaming in English and Spanish for the public and playing prerecorded segments in which we talked about different aspects of the mission. As proud as this moment made me feel, I couldn't help but worry. *Will the rover crash? Will everyone watching live see us fail?* I quickly shook these thoughts away and continued to focus on the present.

We had set up a console in the shed with live visualization of what the operations team was observing on their screens, mainly the data they were receiving from our spacecraft. Because of the signal lag between Mars and Earth, we received confirmation of each event in the timeline through downlinked data about eleven minutes after it had actually occurred. In 2012, *Curiosity* team members experienced the same situation with a seven-minute lag, which they dubbed the "seven minutes of terror." Now here we were, in the thick of our very own "eleven minutes of terror," hoping the data would bring positive news.

As we continued to wait, restlessly, we passed peanuts around, following a long-standing operations-room tradition. It all had started a few decades earlier, during a NASA JPL mission that was going south fast. Everyone paralyzed in front of their screens, receiving an influx of data that indicated several failures, someone suddenly started passing peanuts around. As they all began to anxiously gobble them up, the mission succeeded. And it became a tradition. Yes, even a bunch of NASA engineers and scientists can be a little superstitious.

"We have confirmation of entry interface," said a voice from the operations room.

Entry had been successful—I exhaled momentarily.

"Navigation has confirmed that the parachute has deployed and we're seeing significant deceleration."

The spacecraft was descending through the Martian atmosphere. *Come on, we got this!* I repeated anxiously in my mind.

"*Perseverance* has now slowed to subsonic speed, and the heat shield has been separated."

The rover was now taking photos that would help it choose the best place to land. I lifted my hands to my head and held my breath.

"Sky-crane maneuver has started."

This was it. By now, we knew that our rover had either crashed, shattering years of work in mere seconds, or landed, but we were still waiting with bated breath for the final call.

"Touchdown confirmed. *Perseverance* is safely on the surface of Mars."

Arms shot up to the sky, people embraced, cheered, sighs of relief, tears, excited laughter, shared celebration . . . We did it! At 12:55 p.m. PST our rover had successfully landed on the red planet, and a few minutes later, we had the first image to prove it: our beautiful *Perseverance*'s towering shadow on the rusty-colored surface of Mars.

After the post-landing celebrations and some interviews, I grabbed my orange electric scooter and rode straight home to get some rest, because like the launch, landing was only another beginning. In twelve hours, I had to be next to the testbed replicating the rover's different states to begin the shadowing process we'd been preparing during the last few months. But it was hard to even think of sleep. My family; friends from Puerto Rico, from New York, from Michigan; teachers—they were all texting me. SHPE was posting videos of my segments and interviews on Instagram. Then some of my friends started making fun of my outfit: "Dad vibes." I cracked up. After a few more phone calls, knowing I had a long night ahead of me, I attempted to sleep. By midnight, I was back with the testbed.

The following day, we started receiving the data with video images of *Perseverance*'s entry, descent, and landing. We now had never-before-seen footage of the entire process, the parachute deployment, the sky-crane maneuver that had originated from a napkin doodle, the jet pack that had allowed for a safe landing. We had expected to see fire coming from the thrusters, but propulsion engineers confirmed that the fuel used creates colorless fumes on Mars. We were in awe.

After traveling nearly 300 million miles, *Perseverance* entered the Martian atmosphere at more than 12,000 miles per hour! If it hadn't been safely stowed away inside a heat-resistant shield, it would have otherwise disintegrated like any other meteorite entering the atmosphere at such speed. The surface of the heat shield reached temperatures higher than 1,600°F, hotter than the surface of the sun! Using the shield and cruise stage, we were able to maneuver it like an airplane, but it more closely resembled a skipping rock thrown across the top of a lake. When the spacecraft slowed down to about 1,000 miles per hour, we deployed the 100-pound hypersonic parachute at an altitude of about 7 miles. This parachute slowed it down to about 200 miles per hour. We then had to cut the parachute, release the heat shield—which acted as a camera lens to our landing systems—and rely on its descent rockets. We couldn't let it come down fully to the surface on rockets because the dust plumes could damage our cameras and instruments, so we guided it precisely to the preferred location at Jezero Crater, to 30 meters above it, and lowered the rover safely via cables—the sky-crane maneuver. Upon touchdown, the cables were cut, and the descent stage safely flew away at max throttle to crash far away from the rover.

Meanwhile, the media blitz continued, but the landing wasn't the only thing that caught their eye. Suddenly the world realized what a diverse group of people we were and how we'd all come together to make this happen. It was one of the most powerful outreach moments I had ever experienced. I even received a personal message from Ronnie Nader Bello, the first Ecuadorian astronaut, also born in Guayaquil. "I want to join others in congratulating you, and I'm pleased to see one of us doing what he loves." Awestruck, I continued reading and paused at this line: "You're honoring your mother and her efforts, and I know how happy she is with your accomplishments because your story is similar to my own." Sure, we had the Ecuadorian bond, but what moved me the most was receiving such an affirmation from a seasoned space engineer, who recognized not only my work but also that of Mami. After all, it had always been a team effort.

Around this time, the team also received a Zoom call from President Biden. The mission leads spoke to him from the downlink room, where operations happen, while I was up the hill in the Mars Yard, watching on my screen while testing additional rover capabilities. Getting recognition from the White House that what we had just accomplished through the pandemic was work worth doing was a remarkable moment . . . a small ember in the chaos of those times.

Operating on Mars time was like living in a perpetual state of jet lag. After a few days, there was a cumulative effect that slowly began to change our physical and mental well-being, much like what I've seen happen to sleep-deprived new parents. Every day got heavier and heavier with work. We were all trying to catch sleep whenever we could, be it at three in the afternoon or six in the morning. Burnout was imminent, yet somehow we managed to make it to the end of our campaign successfully. When our mission manager finally gave us the okay, we released the testbeds and went to get some much-needed rest.

By early March, while some of my colleagues took time off to recover, I kept running at full speed and was now starting to shadow the helicopter team. I observed them as they emulated operations, role-playing, obtaining and analyzing data. I knew the rover side of the helicopter, the base station, and was familiar with operations in a broader sense, but now the team needed me to learn the helicopter's flight software and hardware interactions. The clincher: I had mere days to cram six years' worth of knowledge.

Having to keep all these ducks in a row in various teams that had different ways of operating was incredibly taxing. I'd have to stop myself every so often just to recall what group I was working with in the next shift. But my fix-it mentality and relentless desire to see this mission succeed trumped any hint of exhaustion that by this point was desperately pounding on my door.

The downlink process allowed us to determine the health of the spacecraft and whether we were ready to plan the following cycle. The uplink process happened once the downlink team gave the green light to define the set of constraints necessary, if at all, for the next day's activities. While on the rover side, we mostly had people exclusively focused on uplinking data and another group focused on downlinking, with the helicopter, we all had to do both. This meant that one person was doing the job of what ultimately became six people, once the helicopter operations team was fully staffed. Add that to the fact that while most people on the helicopter team were working during that month on a cycle of three weekly shifts, then two days off, I was jumping between the helicopter team and the mechanical downlink side of the rover, working six or seven days straight. All the elements for a perfect storm were in the making. I had not anticipated this additional struggle, and it only added to the accumulated stress I had begun to carry several months earlier. My brain kept jumping from one problem to the next, constantly trying to solve different issues without allowing for a break in my thoughts. Since my buddy Jesse was working on Mars time too, after our third shifts ended at around six in the morning, we'd often go for a hike to clear our minds before hitting our beds for a few hours, only to get up and do it all over again. My friends helped me keep it together as well as possible, and my desire to be a part of achieving these advancements in space kept me motivated, ready to do whatever was required to succeed, regardless of the cost.

On March 17, the team began to seriously discuss how we were going to deploy the helicopter on Mars. A few days later, we performed a successful role-playing test for the rover and the helicopter in the Mars Yard. This was the very last thing we had to do to be prepared for the following step on Mars.

On April 3, 2021, we successfully deployed the helicopter from the rover's underbelly. This was one of the more daunting moments in this journey because now we were communicating with it wirelessly. We no longer could regulate its temperature through the rover's system. *Ingenuity* was on its own, and it had to survive its first Martian night. Like worried parents, we helplessly watched our child fly the nest. Would it be okay? Would it have enough battery life to provide energy to the heaters and not freeze overnight?

At dawn of the next Martian sol, we received data that confirmed *Ingenuity* had survived the night, and we all let out an enormous sigh of relief. As exhausted as I was, the history and significance of what we were about to do was palpable. I had to keep going.

On April 8, we released the blades and proceeded to begin testing the rotors. We started with relatively slow rotations just to confirm they were fully functioning. Once verified, we ramped them up, pushing the rotors to the threshold at which the helicopter would theoretically lift off without allowing it to do so yet. That's how we confirmed we could achieve the necessary speed for it to fly.

The night before the helicopter was scheduled to attempt its first flight, I was on console, so I personally programmed the vehicle's wake-up time for the following Mars morning. On April 19, 2021, *Ingenuity* successfully woke up on the red planet and began to run the sequences that would prepare it to fly. The rotors began to turn and accelerate faster, faster, faster, and then . . . liftoff! It climbed ten feet upward, hovered in the air briefly, rotated, and then safely landed on the surface of Mars. With its first flight complete, *Ingenuity* became the first aircraft in history to make a powered, controlled flight on another planet!

I was so hyped up from the rush of the moment that once my shift was over, even though my body was pleading for rest, I went on a six-mile hike to burn off the adrenaline high. When I reached the top of that mountain, I got on a video call with the helicopter team, and while I spoke to them, I caught sight of a helicopter flying over the city below and thought, *Wait, we just flew a helicopter on Mars!* On one hand, we'd been working so hard on this for so long, and on the other hand, it seemed like just yesterday that I was a little kid playing with toy versions of these machines, breaking them apart to figure out how they worked.

Exhaustion began to take a real hold over me, and insomnia plagued the few hours of rest I had each day, which fed into an intensifying feeling of deep anxiety. But our helicopter's work wasn't over yet. We still had four more scheduled flights to take to call this mission successful. So I powered through, operating more like a machine than a human.

During the helicopter's third flight, it managed to snap a picture of *Perseverance* below, gifting us with yet another unforgettable moment! And then on April 30, my birthday, the fourth flight went off without a hitch, climbing to sixteen feet, flying south, and completing an eight-hundred-seventy-two-foot round trip. *Ingenuity*'s fifth flight happened on May 7, 2021, when it reached a new height record of thirty-three feet and was airborne for close to two minutes. Mission accomplished. We had flown on another planet for the first time, then done it again four more times, and our chopper was still standing!

As a fully functional helicopter on Mars that had passed its technical demo with flying colors, *Ingenuity* then became part of the science mission and would be used to collect more images and scout sites to help us determine whether we should send the rover in a certain direction or not. It is even being used to explore the debris site of the descent stage that helped safely land the rover. And finally, *Ingenuity* is helping scientists choose sites to drop off samples for the eventual Mars Sample Return mission. It recently completed its thirty-fifth flight after surviving the harsh conditions of a Martian winter and even set a new altitude record at forty-six feet.

So many years were devoted to getting *Perseverance* to Mars in one piece and having it survive those first few days on the red planet's surface that once the helicopter completed its mission, there was a definite sense of closure and accomplishment. Now we wait for the Mars 2020 samples to be delivered by the Mars Sample Return mission, in hopes the results will answer the ultimate question, the one that has inspired countless books, movies, documentaries, and dreams: Has life ever existed on Mars?

As for me, at the core of all of the missions, buried under layers of exhaustion and impending burnout, there was another question silently begging to be answered: *What life exists for me here?*

CHAPTER 15

SURFACE OPERATIONS
Searching for Joy

I f humans were to travel through space to Mars one day, it would mean spending months in a cramped capsule, isolated from family and friends, far from Earth and everyday comforts. Ironically, that's how I felt during the last two years of our mission. In 2020, while I navigated personal trials and the pandemic, work became my coping mechanism, much like how I saw my mom operate during her years on the job. As I plunged deeper into my testbed and operations world, and took on more and more responsibilities, the job itself also intensified. Overnight shifts were par for the course, sleep became a luxury attained in whatever free time I could carve out of the never-ending days, and everything I did became mission critical in my mind. I reverted to super-brain fix-it mode, with the red planet as my guiding force. Even though I was running on empty, I went into a machine-like modus operandi, without considering the other parts of me that required attention and fine-tuning to fully function at their best . . . the human parts of me. My body began to give off warning signs, screaming at me to stop, much like Mami's body had signaled her during different stages of her life, and like her, I didn't pay attention until it was too late.

By the time spring of 2021 rolled around, as we marked milestones and made space history, I reached one of the lowest points of my mental health I had ever experienced in my life. My pounding heart would wake me in the middle of the night and propel me into a full-scale fight-or-flight response. I'd stare at the ceiling, trying to reason my way through the very real physical sensations, but the sense of calm that leads to sleep was elusive. The clock was my worst enemy, mocking me with every passing minute, while I counted three . . . two . . . only one hour before I had to wake up for work and go, go, go all over again. Once the crucial phases of our Mars 2020 mission concluded, it was as if I had detached from my body, like a jetpack running out of fuel in midair. My exhaustion was so acute, I approached my lead and finally said, "Hey, I need you to be patient with me for the next few months because I'm not at my best." He didn't know the details but was supportive and understanding. I took some sick days to recover, but that wasn't enough. So I asked for a week off and flew to New York. I needed to ground myself, reconnect with my family, find my way back to me.

"Elio, you look like shit," said Tía Pilar, after greeting me with a big hug, then taking a step back to get a better look at me.

"You know what? Yeah, you're right," I said, laughing nervously.

One would think I should've learned from what happened to Mami with her stroke, heart attack, and repeating panic attacks— explicit reminders of how important balance is to our well-being. But more often than not, we must hit our own walls to be shaken awake. After more than a year of pushing myself way too hard without taking a beat to breathe, I had completely lost my grasp on what life existed for me here, and I finally reached my breaking point.

As immigrants and first-generation kids, most of us have grown up seeing our families work themselves to the bone, sacrificing everything to keep us afloat. There were no vacations, no beach getaways, oftentimes no free weekends—Sundays were usually for church going, cleaning, cooking, and prepping for the next week's daily grind. The only breaks came a few times a year for special occasions and holidays, but those were usually relegated to an afternoon or one day at most. The overriding message was that stopping was not an option. Stopping could mean having everything we've worked so hard to create and accomplish crumble before our eyes. When our families, our parents, our moms sacrifice everything to give us a better life, we become a vital part of what will lead our circle out of poverty. They pass the torch on to us, and we are meant to reach the finish line and at long last attain the American dream. And so we focus solely on mission success and believe we are invincible . . . until we're not. Because we are not robots, we are not machines. We are humans. And as humans, we need to take a breather. Recalculate. Pivot. Find that elusive thing called balance. And as humans, we need to feel. Laugh. Have fun. Risk vulnerability. Show up for ourselves and others. We need to experience the fullness of our humanity.

Throughout the years, my mentors always landed on one piece of common advice: "Don't be so hard on yourself, Elio." I didn't get it. I found it insulting, as if they were saying I couldn't handle my workload. *What the heck is this person talking about? Slow down? Who are they to tell me this?* The message was there, but I wasn't listening. I took on way too much for many years and relied on my managers to tell me when to cut back and to define boundaries that should've been established by me. A part of me felt I had to prove myself, prove that I deserved to be there. Not just to make my family proud but also to show the rest of the world that I was worthy of taking up that space. This is still a challenge for me. Sometimes, when I look at all my achievements, I can't believe I'm in this position. I can't believe that's me in the room with all those incredible minds. I feel like I'm nothing compared to them. Imposter syndrome is real, continuing to be an unwelcome companion on my journey; it's a crazy loop that feeds into what, for me, has become a cycle of anxiety.

I have begun to actively recognize and work on dismantling these thoughts so I can be more present in the moment, which is often hard for someone who has learned how to constantly anticipate and solve problems for a living. I remind myself that those brilliant minds don't treat me differently, as if I'm inferior, that I have earned my spot in this space. I deserve to be here. So the question is not "How am I, of all people, here?" but rather "How do I want to show up here?"

My breaking point and burnout forced me to look inward and redefine my boundaries in my personal and professional lives. I dove into all things healing, and I leaned on my incredible group of friends, the people I knew I could call up anytime and they'd listen and make me laugh. My roommates were fantastic, my mentors offered guidance, and Mami, of course, remained steadfast. But I needed to take it a step further. I couldn't continue to use work as my emotional crutch. Keeping my head down and moving forward with blinders on was no longer a feasible solution. I needed professional help to get me through this crisis. I needed someone who would listen and help me see my exhaustion and emotions for what they were so that I could embark on a journey of healing and self-improvement. Although I knew therapy was an important tool, one that had helped my mom in the past, it still took me a while to take that last step. My main excuse was lack of time, which was just another manifestation of how I didn't prioritize my own well-being. I finally made the call and booked my first round of therapy sessions. A few months went by before I began to climb out of that deep hole and acknowledge everything I had experienced over the last year. It was difficult, but little by little, I felt as if each session opened a valve, releasing some of the pressure that had built up not just over the last year but over a lifetime. I came out of it with a new mission: to nourish my self-awareness and self-love, to make room for myself, to prioritize my well-being. I continue to seek therapy when I need external help.

Finding my center again required time and patience, but it eventually happened. I began to develop more empathy for those around me, and I set out to unlearn unhealthy patterns of behavior that exist in our society at large but also in my family, including my mom. Mami taught me to do better, to be better, to seek excellence. But her actions also taught me to avoid my emotions and turn to work as a coping strategy, which of course is a message affirmed in society and culture. For a while, I was angry at her for not doing something about these detrimental habits, for not prioritizing her well-being, but now I realize I was also angry at myself for repeating this very cycle in my life. The fractures that appeared in our relationship soon became paths to healing, as I was able to empathize with her, understanding that, as humans, we are more nuanced than I previously realized. In turn, I began to offer myself the same compassion and empathy that I was extending to Mami.

I started to reconnect with the activities that made me happy, like group salsa dance classes and surfing. I took a step back from work, planned some time off, and focused on me, what would help my mental health recover, so I could go back to my professional goals in a sounder state of mind. With softer deadlines on the horizon at work, because we were already on Mars doing our thing, and the new work-from-home acceptance that resulted from the pandemic, I was able to start traveling again. I went to New York, to Puerto Rico, to Las Vegas, to Hawaii, to connect with family and friends, to attend music festivals, concerts, several weddings, everything that I knew would make me whole again. And I finally began to feel like a human again, more grounded, open, and calm. So much so that it cleared the way to a new and healthier romantic relationship.

While our missions to Mars may pack more punch and get more airtime, the journey into the unexplored spaces within myself has become equally as meaningful, revealing that on the outside, I am a space mechanic, a technical expert, an engineer who thrives on outreach and innovation, but on the inside, I just want to be a good and loving friend, partner, son, and human. Both the inside and outside components carry the same weight and importance in a human's life. Mission success often requires sacrifice, but the moment we feel like we are relinquishing our control and neglecting what puts a smile on our face and what gives us peace is the moment our self-awareness has to kick in, put our work brain on automatic pilot, and grant us the time and space to investigate the issue, course correcting to find our balance once again.

Now I make sure that the projects I take on, the people I spend my precious time with, the extracurriculars, are all recalibrated as needed for my own well-being. Work is no longer my sole priority. Work must also now align with joy. My love of machines must align with my love of people, myself included.

I'm currently in the testing and iteration phase of Operation Joy, moving pieces of my life in favor of this personal mission that strives for advancement, happiness, and compassion, rather than purely for performance and perfection. It's okay to not be perfect.

There is such a thing as good enough, and we need to embrace that type of compassion in our lives and cut ourselves some slack, because sometimes priorities will be hard to manage, sometimes sacrifice will be required, and sometimes objectives will shift, and that is okay. That is what it means to be human. If I would've told Elio from five years ago that I would eventually decide to leave JPL in search of new adventures, he would've thought I was pulling his leg. Mars had become my life—and it will always be a part of my heart—but I am ready to explore other frontiers.

EPILOGUE
Beaming Toward a Compassionate Future

L eaving NASA's Jet Propulsion Laboratory wasn't easy for me. It was my first work home out of college. It shaped me, pushed me to grow as an engineer and achieve historic results in space, but as I climbed out of my burnout and reprioritized my life, I realized the joy had drained out of me. I was no longer thriving in the technical-achievements-driven culture. My life could no longer revolve solely around the next mission. The time had come to try something new, to seek out other challenges and hopefully land in a place that aligned with my newly discovered needs—ones that prioritized the humans behind the missions in order to start building a more compassionate future.

I'm still shooting for the stars, but now, as I join the Blue Origin team as a systems engineer, my eyes are on the moon. The world is at the cusp of creating what I believe will be a sprawling lunar colony, and more players will be needed to deliver whatever is required to have a sustained human presence on the moon. I'm thrilled to be a part of this enormous endeavor, and on a more mass-produced scale no less, designing and manufacturing space systems that may one day cater to the public. The moon will serve as our testing ground for a long-term human presence in space, which will likely help pave the way to eventually landing humans on Mars and beyond.

Having commercial players that can provide launch vehicles is changing the space landscape and making it more accessible. Underserved communities and countries, small companies, universities, and civil agencies around the world will soon have the opportunity to start their own space programs, and use them to expand their communication infrastructure, their education access, and insights into their respective environments. Most modern-day problems have very technical solutions. If approached with human-centered empathy, we could solve a lot of world problems through engineering.

As we progress with these missions, the human experience will expand and we will see advancements in manufacturing, robotics, medicine, agriculture, and space beyond what our minds can currently dream of. Take our Mars 2020 mission. The samples NASA is collecting with *Perseverance* will return to Earth eventually, as another mission is scheduled to head to Mars in the next decade to pick them up and bring them home. In time, *Perseverance* and *Ingenuity* will become antiquated mechanisms that will be surpassed by what future generations will create, build, test, and fly. Who knows—maybe one day we will find a way to send a series of submarine drones and drilling robots to Europa, Jupiter's fourth largest moon, to pierce its ice shell, which is close to fifteen miles thick, and explore the chemistry of that water. Would we find life? Maybe. Will Mars or Europa help us finally answer the intrinsic philosophical question "Are we alone?" Hopefully.

This is why I believe the world would benefit from having more engineers, but for that to happen, we desperately need to expand access to education. My academic journey absolutely changed the course of my life; that's why I advocate for education with such passion every chance I get and am even considering pursuing new degrees in the not-so-distant future. I have learned, from my own experience, that education truly is a means of escaping the generational suffering that has held people back for years. For those of you who have been told, "Forget school—that doesn't do anything for you. You need to be contributing to your family instead," remember this: As difficult as it may be, there are times when the only way to break the generational poverty cycle is to put your family to the side and go get prepared, get the education that will open the doors you and your family have always deserved. Your family will reap the benefits of your success. With access to education comes access to networks of people, and within those networks we can oftentimes enact direct change for our communities and improve people's lives.

The current and continuous development of technology—like artificial intelligence, advanced manufacturing, augmented and virtual reality, and alternative power generation and collection methods—happening at such an accelerated pace is in need of bright young minds that will continue the limitless work we are creating. For this reason, I believe outreach is imperative and must continue to grow. I recently participated in a talk for Hispanic high school students in a library in North Carolina. I walked into the room and was blown away to find around a hundred students filling every seat, who then eagerly listened to what we had to say, absorbing the possibilities of STEM, the light in their eyes shining brighter as they envisioned a path forward for themselves. I couldn't help but think of my first Noche de Ciencias at that Detroit high school. As a hopeful college freshman, I had been expecting fifty or so attendants and only ten showed up. To go from that small turnout to this library full of eager and engaged Hispanic high school students was an enormous stride I could barely imagine possible at the time. Nevertheless, both events carried equal importance because each individual life carries equal importance, and all it takes is the ability to inspire one life to change many for the better. Both of those events could result in one more Hispanic joining the field and changing the tapestry of STEM so that we can eventually reflect our actual population.

Hispanics make up around 19 percent of the nation's population, yet according to a Pew Research Center report published in April 2021, we represent only 8 percent of STEM workers and just a measly 5 percent of the engineering workforce. Strides are being taken, diversity and representation are increasing, yet we still have a long way to go. I hope my story inspires people to change these statistics. The space industry is open for the taking, and if after reading this book even one person is motivated to pursue an engineering degree and join me on the journey of pioneering space, then one of my missions on Earth will be closer to completion.

Space is an incredible frontier that is helping us develop technology, expand our knowledge, and understand our position in the universe. Being able to look up and understand that we are a tiny dust particle in the universe has helped me develop humility and a deeper appreciation for our planet Earth. Now I want to cross the Kármán line and experience the cognitive change of consciousness called the "overview effect," which comes from observing our planet from a completely new perspective. So as I stand on the shoulders of giants like Ellen Ochoa, José Hernández, and Franklin Chang-Díaz—three of thirteen Hispanics who have been to space—I've applied for

Space for Humanity's first sponsored Citizen Astronaut Program. Being accepted into this program would allow me to represent my people, culture, music, food, and language as we continue to establish presence beyond the limits of Earth.

If this first attempt to reach space doesn't pan out, I will find another way. Because I'm determined to be a part of paving this new way forward, so that children can see people who look like them taking on extraordinary feats, so that my grandparents can see their family cross new frontiers, and so that my mom can rise with me, ascending to the stars through me.

I want to be a part of a universe where the echoes of Latin America continue to reach the stars.

From that height, it becomes ever so clear that we are such a vulnerable, unique, tiny, beautiful blue dot in the universe. So look up, feel your feet, and understand that it's up to all of us to take care of the ground below us, the environment around us, the humans next to and within us, the air we breathe in Earth's atmosphere, and even the space above us. No matter what mission you embark on next, persevere, be ingenious, take the opportunity, bring others along for the climb, remain compassionate with yourself and those around you, and always keep reaching for the stars.

ACKNOWLEDGMENTS

Thank you, Mami, for sacrificing so much for my well-being. My multiplanetary adventure would not be possible without your work ethic, warmth, wisdom, and guidance. Your love for me has always provided constant safety and reassurance. Thank you for being the strongest and most beautiful person I know. I owe you everything.

Thank you to my chosen family, from Caguas, Puerto Rico, to Brooklyn, New York; from Ann Arbor, Michigan, to Los Angeles, California. Even in the depths of freezing temperatures, I have always been wrapped in love and warmth by some of the most beautiful people in the world. My LA homies, my SHPE familia, my friends that love snacks, my Nesties, my Notre crew, this is all thanks to your unconditional support.

To Johanna Castillo, my literary agent, for reaching out to me and passionately guiding me through this new world.

To Cecilia Molinari, for being an incredible listener and putting my story into a beautiful narrative, but especially for becoming such a wonderful friend.

To Judith Curr and the incredible team at HarperOne, especially Juan Mila and Noelle Olmstead, for their careful edits and support to make this book a reality.

To mis abuelos, whose nearly seventy-five years of marriage exemplify true love.

To my tíos and tías, Oscar, Jenny, Vicky, Eli, Miriam, and Lucho, for showing me support through the years even if at a distance.

To my brother, Xavier, for showing me incredible resilience.

To my nieces, Amber and Arianne, for being the best pains in my butt I could ever ask for.

To Sonia, Robert, Benito, Gabriela: thank you for also being my family.

To my teachers and mentors, especially Darryl Koch for opening the doors to the University of Michigan. Thank you for creating an academic journey that has led me to space exploration.

To my JPLers, especially David Henriquez, Eric Aguilar, and Magdy Bareh, for paving the path to the Mars program.

To my peers and all of those who came before me in the space industry, thank you. Here's to inspiring those who come next and building a future where the Earth can thrive while we also venture to the stars. Sí se pudo, sí se puede, sí se podrá.

REFERENCES

Following is a list of resources to help further expand your own space knowledge based on what we've explored together in this journey to the stars.

Black Holes: https://quantumfrontiers.com/2014/06/20/ten-reasons -why-black-holes-exist

https://exoplanets.nasa.gov/resources/2330/collection-of-interactive -powerpoint-slides-to-be-used-in-public-engagement

Blue Origin: https://www.blueorigin.com

***Challenger* Shuttle Tragedy:** https://www.space.com/18084-space -shuttle-challenger.html

DART (Double Asteroid Redirection Test): https://dart.jhuapl.edu /News-and-Resources/index.php

https://solarsystem.nasa.gov/missions/dart/in-depth

Documentary on *Opportunity*: *Good Night Oppy*, streaming on Amazon Prime Video.

REFERENCES

Exoplanets: https://exoplanets.nasa.gov/trappist1

https://exoplanets.nasa.gov/search-for-life/habitable-zone/

JPL (Jet Propulsion Laboratory): https://www.jpl.nasa.gov

Landing on Mars: https://theconversation.com/decades-of-attempts
-show-how-hard-it-is-to-land-on-mars-heres-how-we-plan-to-succeed
-in-2021-69734

Mars Sample Return: https://mars.nasa.gov/msr

NASA's Hispanic Astronauts: https://www.nasa.gov/sites/default/files
/atoms/files/hispanic_astronauts_fs.pdf

NASA *Perseverance*: https://mars.nasa.gov/mars2020

Notable Hispanic or Hispanic-Descendant Astronauts:

Arnaldo Tamayo Mendez (Cuba): First person from Latin America and
first Black person in space. Flew on Soyuz 38 (September 18, 1980).
https://www.britannica.com/biography/Arnaldo-Tamayo-Mendez

Rodolfo Neri Vela (Mexico): First Mexican in space. Flew on STS-61-B
(November 26, 1985). https://www.britannica.com/biography/Rodolfo
-Neri-Vela

Ellen Ochoa (United States, Mexican-descendent): First Hispanic
woman in space. First flew on STS-56 (April 8, 1993). https://www
.nasa.gov/sites/default/files/atoms/files/ochoa.pdf

Sidney M. Gutierrez (United States): First US-born Hispanic astronaut.
First flew on STS-40 (June 5, 1991). https://www.nmspacemuseum
.org/inductee/sidney-m-gutierrez/?doing_wp_cron=1675346146
.6304459571838378906250

Joseph M. Acaba (Puerto Rico): First Puerto Rican in space. First flew on STS-119 (March 15, 2009). https://www.nasa.gov/astronauts /biographies/joseph-m-acaba/biography

Katya Echazarreta (United States, Mexican-born): First Mexican woman in space. Flew on Blue Origin NS-21 (June 4, 2022). https:// hnmagazine.com/2022/06/meet-katya-echazarreta-first-mexican -born-woman-go-space/

Orbits: https://www.nasa.gov/audience/forstudents/5-8/features/nasa -knows/what-is-orbit-58.html

Perseverance **Findings:** https://scitechdaily.com/life-on-mars-latest -intriguing-organic-findings-by-nasas-perseverance-rover

Skycrane Maneuver: https://astronomy.com/news/2021/02/skycrane -how-perseverance-will-land-on-mars

Space Programs in Latin America: History, Current Operations, and Future Cooperation: https://www.airuniversity.af.edu/Portals/10/JOTA /Journals/Volume%203%20Issue%203/04-Guzman_eng.pdf

Voyager **in Depth:** https://solarsystem.nasa.gov/missions/voyager-2/in -depth

Why NASA's Interstellar Mission Almost Didn't Happen: https://www .nationalgeographic.com/magazine/article/explore-space-voyager -spacecraft-turns-40